W0115170

# Nested Partitions Method, Theory and Applications

**\*A list of the early publications in the series is at the end of the book\***

# Nested Partitions Method, Theory and Applications

**Leyuan Shi**
University of Wisconsin-Madison, WI, USA

**Sigurdur Ólafsson**
Iowa State University, IA, USA

 Springer

Leyuan Shi
University of Wisconsin-Madison
WI, USA
leyuan@engr.wisc.edu

Sigurdur Ólafsson
Iowa State University
IA, USA
olafsson@iastate.edu

ISBN: 978-0-387-71908-5         e-ISBN: 978-0-387-71909-2

Library of Congress Control Number: 2008934910

© Springer Science+Business Media, LLC 2009
All rights reserved. This work may not be translated or copied in whole or in part without the written permission of the publisher (Springer Science+Business Media, LLC, 233 Spring Street, New York, NY 10013, USA), except for brief excerpts in connection with reviews or scholarly analysis. Use in connection with any form of information storage and retrieval, electronic adaptation, computer software, or by similar or dissimilar methodology now know or hereafter developed is forbidden.
The use in this publication of trade names, trademarks, service marks and similar terms, even if they are not identified as such, is not to be taken as an expression of opinion as to whether or not they are subject to proprietary rights.

Printed on acid-free paper

springer.com

To my parents Dangping Shi and Wanrong Shen
*LS*

To ömmu Lilju, mömmu, and Jenny
*SÓ*

# Contents

# 1

# Introduction

The subject of this book is the *nested partitions method* (NP), a relatively new optimization method that has been found to be very effective solving discrete optimization problems. Such discrete problems are common in many practical applications and the NP method is thus useful in diverse application areas. It can be applied to both operational and planning problems and has been demonstrated to effectively solve complex problems in both manufacturing and service industries. To illustrate its broad applicability and effectiveness, in this book we will show how the NP method has been successful in solving complex problems in planning and scheduling, logistics and transportation, supply chain design, data mining, and health care. All of these diverse applications have one characteristic in common: they all lead to complex large-scale discrete optimization problems that are intractable using traditional optimization methods.

## 1.1 Large-Scale Optimization

In developing the NP method we will consider optimization problems that can be stated mathematically in the following generic form:

$$\min_{x \in X} f(x), \tag{1.1}$$

where the solution space or feasible region $X$ is either a discrete or bounded set of feasible solutions. We denote a solution to this problem $x^*$ and the objective function value $f^* = f(x^*)$.

For most of the applications considered in this book the feasible region $X$ is finite but its size grows exponentially in the input parameters of the problem. In many cases $X$ also has complicated constraints that are difficult to satisfy. The objective function $f: X \rightarrow \mathbf{R}$ is usually a complex non-linear function. Sometimes it may have no analytic expression and must be evaluated through a model, such as a simulation model, a data mining model, or other application-dependent models. One advantage of the NP method is that it is effective for optimization when $f$ is known analytically (deterministic optimization), when it is noisy (stochastic optimization), or even when it must be evaluated using an external process.

### 1.1.1 Exact Solution Methods

The need to solve discrete optimization problems has long been established and has been the subject of intense research (Nemhauser and Wolsey 1988, Schrijver 2005). Such problems are usually formulated as either combinatorial optimization problems (COP) where the feasible region is finite, integer programming (IP) problems where both the constraints and objective function are linear, or mixed integer programming (MIP) problems where some of the variables are discrete and some are continuous. Discrete optimization problems can be addressed using one of three approaches: exact methods, approximation algorithms, or heuristics. Exact methods guarantee the optimal solutions, while approximation algorithms guarantee a solution within a certain distance from the optimum, and heuristics simply seek good solutions without making a performance guarantee.

Exact solution methods are grounded in mathematical programming theory. Such methods have been studied for decades and the last twenty years have seen significant breakthroughs in the ability to solve large-scale discrete problems using mathematical programming. These methods have in recent years been used to solve very large practical problems (Atamturk and Savelsbergh 2005), but from our perspective they require significant structure. For example, the objective function and constraints are typically assumed linear and even then additional structure is usually required for mathematical programming methods to be effective. However, when such structure is present and appropriately assumed, mathematical programming methods are often not only effective but also very efficient. The two primary classes of methods that can be used to solve discrete problems are branching methods such as the classic branch-and-bound (Balas and Toth 1995, Beale 1979) and the more recent branch-and-cut (Balas et al. 1996, Caprara and Fishetti 1997, Martin 2001, Padberg 2005), and decomposition methods such as Lagrangian relaxation (LR) (Beasley 1993, Frangioni 2005) and Dantzig-Wolfe (DW) decomposition (Villeneuve et al. 2005, Vanderbeck and Savelsbergh 2006).

Branching methods divide the solution space into partitions called branches and focus the computational effort on obtaining tight lower bounds $\underline{f} \leq f^*$ for each branch. They then use these bounds to eliminate branches where the lower bound is worse than some known feasible solution $x^0 \in X$, that is, $f(x^0) < \underline{f}$. In this manner all the feasible solutions can often be accounted for by considering relatively few branches. This does, however, rely on the ability to obtain tight lower bounds. These bounds can be found by solving simple relaxations, such as linear programming relaxation of an MIP, or improved by generating inequalities that cut off non-feasible solutions (branch-and-cut). Although branching methods have been applied successfully for many problems, it may be very difficult or impossible to find sufficiently good bounds for complex problems.

Decomposition methods solve the problem by either eliminating constraints (e.g., Lagrangian relaxation) or variables (e.g., Dantzig-Wolfe decomposition). Branching and decomposition can also be combined, as is done for example by the branch-and-price algorithm or by using LR to find tighter bounds in branch-and-bound. Unfortunately exact solution methods cannot usually be applied effectively to large-scale optimization unless the structure of the problem is relatively simple, such as when the objective function and all of the constraints are linear.

Approximation algorithms are often based on similar mathematical programming principles as the exact algorithms. For example, say that a relaxed problem has been solved for a lower bound $\underline{f}$ and a feasible solution $x^0 \in X$ has been found. Then it can be said that $x^0$ is an $\epsilon$-approximation if $f\left(x^0\right) - \underline{f} \le \epsilon$.

### 1.1.2 Heuristic Solution Methods

When exact methods cannot find the optimal solution (or even a sufficiently good approximation) in a reasonable amount of time, heuristics must be relied on to find good solutions (Reeves, 1993; Smith, 1996). While heuristics do not guarantee the performance they have the advantage that they do not make restrictive assumptions about the structure of the problem and are therefore applicable to a wider range of applications. They can often quickly generate very good solutions, which is critical when the time to solve the problem is limited.

From an optimization theory point of view there is a significant philosophical shift when applying heuristics rather than exact methods. To guarantee performance, most exact methods focus the majority of the computational effort on obtaining tight lower bounds. When these bounds become sufficiently tight, a feasible solution is generated that, due to the bounds, is then shown to be optimal. However, only a small fraction of the computational effort is devoted to generating feasible solutions. Heuristics, on the other hand, devote most or all of the computational effort to generating a sequence $x^0, x^1, ..., x^k$ of feasible solutions. Continuing with the bounding perspective of exact methods, this does of course now provide an upper bound

$$\bar{f} = \min_i f\left(x^i\right) \ge f^*$$

on the performance.

The simplest type of a heuristic is a greedy local search, where in each iteration a move is made that improves performance, that is, $f\left(x^0\right) > f\left(x^1\right) > ... > f\left(x^k\right) \ge f^*$. A move is defined as some small change to the current solution and is application-dependent. A greedy search is typically very fast and terminates when no further improvement is possible. However, for all but the simplest problems there are multiple local optima and the gap $f\left(x^k\right) - f^*$ is usually unsatisfactorily large because the greedy search becomes stuck at the first local optimum that is encountered. This has led to the development of many randomized heuristics and metaheuristics (Gendreau and Potvin 2005, Glover and Kochenberger 2003, Lovsz, 1996).

A random search adds an element that enables a greedy local search to escape from local optima. For example, simulated annealing allows a move to an inferior solution with a probability that depends on the gap between the performance of the current solution and the inferior solution, and a parameter called the temperature (Kirkpatrick et al. 1983, Eglese 1990, Fleischer 1995). For a candidate new solution, if the difference in performance is small and the temperature is high then it is likely that the candidate is accepted and becomes the new solution in the sequence. Vice versa, if the difference is large and the temperature is low, it is unlikely that this candidate will be accepted. The algorithm terminates by systematically decreasing the temperature parameter, which eventually results in a local optimum that cannot be escaped. Given certain conditions, it can be assured that this final local optimum is also a global optimum and convergence is asymptotically guaranteed. Many other heuristics have been suggested and found useful, including genetic algorithms and other evolutionary methods (Goldberg 1989, Leipins and Hillard 1989, Muhlenbien 1997), tabu search (Glover 1989, Glover 1990, Glover and Laguna, 1997), scatter search (Glover 1997, Glover, Laguna and Marti 2003), variable neighborhood search (Hansen and Mladenovic 1997), and ant-colony optimization (Dorigo and Stutzle, 2004). In the next section we place the NP method in the context of such metaheuristics.

## 1.2 The NP method

The NP method is best viewed as a metaheuristic framework and it has similarities to branching methods in that like branch-and-bound it creates partitions of the feasible region. However, it also has some unique features that make it well-suited for very difficult large-scale optimization problems.

Metaheuristics have emerged as the most widely used approach for solving difficult large-scale combinatorial optimization problems. A metaheuristic provides a framework for guiding application-specific heuristics, such as a greedy local search, by restricting which solution or set of solutions should or can be visited next. For example, the tabu search metaheuristic disallows certain moves that might otherwise be appealing by making the reverse of recent moves tabu or forbidden. At the same time it always forces the search to take the best non-tabu move, which enables the search to escape local optima. Similar to tabu search, most metaheuristics guide the search from solution to solution or possibly from one set of solutions to another set of solutions. In contrast, the NP method guides the search by determining where to concentrate the search effort. Any optimization method, such as an application-specific local search, general purpose heuristic, or a mathematical programming method, can then be integrated within this framework.

Metaheuristics and other heuristic search methods have been developed largely in isolation from the recent advances in the use of mathematical programming methods for solving large-scale discrete problems. It is a very important and novel characteristic of the NP method that it provides a natural metaheuristic framework for combining the use of heuristics and mathematical programming and for taking advantage of their complimentary nature. Indeed, as far as we know, the NP method is the first systematic search method that enables users to simultaneously realize the full benefits of incorporating lower bounds through various mathematical programming methods and using any domain knowledge or heuristic search method for generating good feasible solutions. It is this flexibility that makes the NP method so effective for practical problems.

To concentrate the search effort the NP method employs a decomposition approach similar to that of branch-and-bound. Specifically, in each step the method partitions the space of feasible solutions into the *most promising region* and the *complimentary region*, namely the set of solutions not contained in the most promising region. The most promising region is then further partitioned into subregions. The partitioning can be done exactly as branching for a branch-and-bound algorithm would be, but instead of focusing on obtaining lower bounds and comparing those bounds to a single primal feasible solution, the NP method focuses on generating primal feasible solutions from each of the subregions and the complimentary region. This results in an upper bound on the performance of each of these regions. The region with the best feasible solution is judged the most promising and the search focused accordingly. A best upper bound does not guarantee that the corresponding subset contains the optimal solution, but since the NP method also finds primal feasible solutions for the complimentary region, it is able to recover from incorrect moves. Specifically, if the best solution is found in one of the subregion,s this becomes the new most promising region, whereas if it is in the complimentary region the NP method backtracks. This focus on generating primal feasible solutions and the global perspective it achieves through backtracking are distinguishing features of the NP method that set it apart from similar branching methods.

Unlike exact optimization methods such as branch-and-bound, the NP method does not guarantee that the correct region is selected in each move of the algorithm. Incorrect moves can be corrected through backtracking, but for the method to be both effective and efficient, the correct move must be made frequently. How this is accomplished depends on how the feasible solutions are generated.

In what we will refer to as the *pure NP method*, feasible solutions are generated using simple uniform random sampling. To increase the probability of making the correct move, the number of samples should be increased. We will later see how statistical selection methods can be used to prescribe a sufficient amount of sampling in order to assure that the correct region is selected with a given probability. A purely uniform random sampling is rarely efficient, however, and the strength of the NP method is that it can incorporate application-specific methods for generating feasible solutions. In particular, for practical applications domain knowledge can often be utilized to very effectively generate good feasible solutions. We call such implementations *knowledge-based NP methods*. We will also see examples of what we refer to as *hybrid NP methods* where feasible solutions are generated using either general heuristic methods such as greedy local search, genetic algorithms or tabu search, or mathematical programming methods. If done effectively, incorporating such methods into the NP framework makes it more likely that the correct move is made and hence makes the NP method more efficient. Indeed, we will see that such hybrid and knowledge-based implementations are often an order of magnitude more efficient than uniform random sampling.

In addition to the method for generating feasible solutions, the probability of making the correct move depends heavily on the partitioning approach. The implementation of a generic method for partitioning is usually straightforward but by taking advantage of special structure and incorporating this into intelligent partitioning, the efficiency of the NP method may be improved by an order of magnitude. The strength of the NP method lies indeed in this flexibility. Special structure, local search, any heuristic search, and mathematical programming can all be incorporated into the NP framework to develop optimization algorithms that are more effective in solving large-scale optimization problems than when these methods are used alone.

## 1.3 Application Examples

In this section we introduce three application examples that illustrate the type of optimization problems for which the NP method is particularly effective. For each application the optimization problem has a complicating aspect that makes it difficult for traditional optimization methods. For the first of these problems, resource-constrained project scheduling, the primary difficulty lies in a set of complicating constraints. For the second problem, the feature selection problem, the difficulty lies in a complex objective function. The third problem, radiation treatment planning, has constraints that are difficult to satisfy as well as a complex objective function that cannot be evaluated through an analytical expression. Each of the three problems can be effectively solved by the NP method by incorporating our understanding of the application into the framework.

### 1.3.1 Resource-Constrained Project Scheduling

Planning and scheduling problems arise as critical challenges in many manufacturing and service applications. One such problem is the resource-constrained project scheduling problem that can be described as follows (Herroelen and Demeulemeester 1994). A project consists of a set of tasks to be performed and given precedence requirements between some of the tasks. The project scheduling problem involves finding the starting time of each task so that the overall completion time of the project is minimized. It is well known that this problem can be solved efficiently using what is called the critical path method that uses forward recursion to find the earliest possible completion time for each task (Pinedo 2000). The completion time of the last task defines the makespan or the completion time of the entire project.

Now assume that one or more resource is required to complete each task. The resources are limited so if a set of tasks requires more than the available resources they cannot be performed concurrently. The problem now becomes NP-hard and cannot be solved efficiently to optimality using any traditional methods. To state the problem we need the following notation:

$$V = \text{Set of all tasks}$$
$$E = \text{Set of precedence constraints}$$
$$p_i = \text{Processing time of task } i \in V$$
$$R = \text{Set of resources}$$
$$R_k = \text{Available resources of type } k \in R.$$
$$r_{ik} = \text{Resources of type } k \text{ required by task } i.$$

The decision variables are the starting times for each task,

$$x_i = \text{Starting time of task } i \in V \tag{1.2}$$

Finally, for notational convenience we define the set of tasks processed at time $t$ as

$$V(t) = \{i : x_i \leq t \leq x_i + p_i\}.$$

With this notation, we now formulate the resource-constrained project scheduling problem mathematically as follows:

$$\min \max_{i \in V} x_i + p_i \tag{1.3}$$

$$x_i + p_i \leq x_j, \quad \forall (i,j) \in E \tag{1.4}$$

$$\sum_{i \in V(t)} r_{ik} \leq R_k, \quad \forall k \in R, t \in \mathbf{Z}_+^1 \tag{1.5}$$

$$x_i \in \mathbf{Z}_+^1$$

Here the precedence constraints (1.4) are easy, whereas the resource constraints (1.5) are hard. By this we mean that if the constraints (1.5) are dropped then the problem becomes easy to solve. Such problems, where complicating constraints transform the problem from easy to very hard, are common in large-scale optimization. Indeed the classic job shop scheduling problem can be viewed as a special case of the resource constrained project scheduling problem where the machines are the resources. Without the machine availability constraints the job shop scheduling problem reduces to a simple project scheduling problem. Other well-known combinatorial optimization problems have similar properties. For example, without the subset elimination constraints the classic traveling salesman problem reduces to a simple assignment problem that can be efficiently solved.

The flexibility of the NP method allows us to effectively address such problems by taking advantage of special structure when generating feasible solutions. It is important to note that it is very easy to use sampling to generate feasible solutions that satisfy very complicated constraints, which are very difficult to handle using traditional methods such as mathematical programming. Therefore, when faced with a problem with complicating constraints we want to use random sampling to generate partial feasible solutions that resolve the difficult part of the problem and then complete the solution using the appropriate efficient optimization method.

For example, when generating a feasible solution for the resource constrained project scheduling problem, the resource allocation should be generated using random sampling and the solution can then be completed by applying the critical path method to determine the starting times for each task. This requires reformulating the problem so that the resource and precedence constraints can be separated. Such a reformulation is rather easily achieved by noting that the resource constraints can be resolved by determining a sequence between the tasks that require the same resource(s) at the the same time. Once this sequence is determined then the sequence can be added as a set of precedence constraints, which are easy to deal with, and the remaining solution can be generated using the critical path method. Feasible solutions can therefore be generated in the NP method by first randomly sampling a sequence to resolve resource conflicts and then applying the critical path method. Both procedures are very fast, so complete sample solutions can be generated rapidly.

We also note that constraints that are difficult for optimization methods such as mathematical programming are sometime very easily addressed in practice by incorporating domain knowledge. For example, a domain expert may easily be able to specify priorities among tasks requiring the same resource(s) in the resource-constrained project scheduling problem. The domain expert can therefore, perhaps with some assistance from an interactive decision support system, specify some priority rules to convert a very complex problem into an easily solved problem. The NP method can effectively incorporate such domain knowledge into the optimization framework by using the priority rules when generating feasible solutions. This is particularly effective because the domain expert would not need to specify priority rules to resolve all resource conflicts. Rather, any available priority rule or other domain knowledge can be incorporated to guide the sampling.

The same structure can be used to partition intelligently. Instead of partitioning directly using the decision variables (1.2), we note that it is sufficient to partition to resolve the resource conflicts. Once those are resolved then the problem is solved. This approach is applicable to any problem that can be decomposed in a similar manner.

We will revisit the resource-constrained project scheduling problem in both Chapter 4 and Chapter 5, where we discuss further how to incorporate mathematical programming techniques and domain knowledge, respectively, to solve this problem more efficiently.

### 1.3.2 Feature Selection

Knowledge discovery and data mining is a relatively new field that has experienced rapid growth due to its ability to extract meaningful knowledge from very large databases. One of the problems that must usually be solved as part of practical data mining projects is the feature selection problem (Liu and Motoda 1998), which involves selecting a good subset of variables to be used by subsequent inductive data mining algorithms. The problem of selecting the best subset of variables is well known in the statistical literature as well as in machine learning. The recent explosion of interest in data mining for addressing various business problems has led to a renewed interest in this problem. From an optimization point of view, feature selection can clearly be formulated as a binary combinatorial optimization problem where the decision variables determine whether a feature (variable) is included or excluded. The solution space can therefore be stated very simply as all permutations of a binary vector of length $n$, where $n$ is the number of variables. The size of this feasible region is $2^n$ so it experiences exponential growth, but typically there are no additional constraints to complicate its structure.

On the other hand, there is no consensus objective function that measures the quality of a feature or a set of features. Tens of alternatives have been proposed in the literature, including functions that measure the quality of individual features as well as those that measure the quality of a set of features. However, no single measure is satisfactory in all cases and the ultimate measure is therefore: does it work? In other words, when the selected features are used for learning, does this result in a good model being induced? The most effective feature selection approach in terms of solution quality is therefore the wrapper approach, where the quality of a set of features is evaluated by applying a learning algorithm to the set and evaluating its performance. Specifically, an inductive learning algorithm, such as decision tree induction, support vector machines or neural networks, are applied to a training data set containing only the selected features. The performance of the induced model is evaluated and this performance is used to measure the quality of the feature subset. This objective function is not only non-linear, but since a new model must be induced for every feature subset it is very expensive to evaluate.

Mathematically, the feature selection problem can be stated as follows:

$$\min_{x \in \{0,1\}^n} f(x),  \tag{1.6}$$

that is, $X = \{0,1\}^n$. Feature selection is therefore a very difficult combinatorial optimization problem not because of the complexity of the feasible region (although it does grow exponentially), but due to the great complexity of an objective function that is very expensive to evaluate. However, this is also an example where application-specific heuristics can be effectively exploited by the NP method.

As previously stated, significant research has been devoted to methods for measuring the quality of features. This includes information-theoretic methods such as using Shannon's entropy to measure the amount of information contained in each feature: the more information the more valuable the feature. The entropy is measured individually for each feature and can thus be used as a very fast local search or a greedy heuristic, where the features with the highest information gain are added one at a time. While such a purely entropy-based feature selection will rarely lead to satisfactory results, the NP method can exploit this by using the entropy measure to define an intelligent partitioning that is an order of magnitude more efficient than an average arbitrary partitioning. It can also be used to generate feasible solutions from each region using a sampling strategy that is biased towards including features with high information. A very fast greedy heuristic can thus greatly increase the efficiency of the NP method while resulting in much higher-quality solutions that the greedy heuristic is not able to achieve on its own. This feature selection problem is explored further in Chapter 7.

### 1.3.3 Radiation Treatment Planning

Health care delivery is an area of immense importance where optimization techniques have been used increasingly in recent years. Radiation treatment planning is an important example of this and Intensity-Modulated Radiation Therapy (IMRT) is a recently developed complex technology for such treatment (Lee, Fox and Crocker 2003). It employs a multileaf collimator to shape the beam and to control, or modulate, the amount of radiation that is delivered from each of the delivery directions (relative to the patient). The planning of the IMRT is very important because it needs to achieve the treatment goal while incurring the minimum possible damage to other organs. Because of its complexity the treatment planning problem is generally divided into several subproblems. The first of these is termed the *beam angle selection* (BAS) problem (Djajaputra et al. 2003, D'Souza, Meyer and Shi 2004). In essence, beam angle selection requires the determination of roughly 4-9 angles from 360 possible angles subject to various spacing and opposition constraints.

Designing an optimal IMRT plan requires the selection of beam orientations from which radiation is delivered to the patient. These orientations, called beam angles, are currently manually selected by a clinician based on his/her judgment. The planning process proceeds as follows: a dosimetrist selects a collection of angles and waits ten to thirty minutes while a dose pattern is calculated. The resulting treatment is likely to be unacceptable, so the angles and dose constraints are adjusted, and the process repeats. Finding a suitable collection of angles often takes several hours. The goal of using optimization methods to identify quality angles is to provide a better decision support system to replace the tedious repetitive process just described. An integer programming model of the problem contains a large number of binary variables and the objective value of a feasible point is evaluated by solving a large, continuous optimization problem. For example, in selecting 5 to 10 angles, there are between $4.9^{10}$ and $8.9 \times 10^{19}$ subsets of $0, 1, 2, ..., 359$.

The BAS problem is complicated by both an objective function with no analytical expression and by constraints that are hard to satisfy. In the end an IMRT plan is either acceptable or not and the considerations for determining acceptability are too complex for a simple analytical model. Thus, the acceptability and hence the objective function value for each plan must be evaluated by a qualified physician. This makes evaluating the objective not only expensive in terms of time and effort, but also introduces noise into the objective function because two physicians may not agree on the acceptability of a particular plan. The constraints of the BAS problem are also complicated since each beam angle will result in radiation of organs that are not the target of the treatment. There are therefore two types of constraints: the target should receive at least a minimum radiation and other organs should receive no more than some maximum radiation. Since these bounds need to be specified tightly the constraints are hard to satisfy.

The BAS problem illustrates how mathematical programming can be effectively incorporated into the NP framework. Since the evaluation of even a single IMRT plan must be done by an expert and is thus both time consuming and expensive, it is imperative to impose a good structure on the search space that reduces the number of feasible solutions that need to be generated. This can be accomplished by means of an intelligent partitioning and we do this by computing the optimal solution of an integer program with a much simplified objective function. The output of the IP then serves to define an intelligent partitioning . For example, suppose a good angle set $(50°, 80°, 110°, 250°, 280°, 310°, 350°)$ is found by solving the IP. We can then partition on the first angle in the set, which is $50°$ in this example. Then one sub-region includes angle $50°$, the other excludes $50°$. This partitioning has been found to be very effective and this problem is explored further in Chapter 9.

These three application examples illustrate the broad usefulness of the NP method in both manufacturing and service industries, and how it can take advantage of special structure and application-specific heuristics to improve the efficiency of the search. In Part II of the book we will consider numerous applications in much more detail and show how domain knowledge, greedy heuristics, generic heuristics, and mathematical programming can all be incorporated into the NP framework to efficiently solve very complex problems that arise in a wide range of applications.

## 1.4 About the Book

This book is divided into two parts. Part I develops the general theory and implementation framework for the NP method, while Part II provides a detailed look at several application areas and how the NP method is effectively applied in these areas. Readers who are primarily interested in applications may want to focus on Part II but should first familiarize themselves with the first five sections of Chapter 2, which develop the foundation for the NP method and its implementation. In the remainder of this section we provide a brief overview of each chapter.

The groundwork for the NP method is laid in Chapter 2. Section 2.1 shows a generic implementation of the algorithm and each step is then explored in more detail in the next four sections. Section 2.2 looks at how partitioning is used to impose structure on difficult problems and how intelligent partitioning improves the efficiency of the NP method. Section 2.3 discusses how to effectively generate feasible solutions that are used to guide the search in the NP method. Both generic and application-specific approaches are explored and it is demonstrated that by incorporating domain knowledge, it is possible to greatly improve the efficiency of the method. Two additional implementation issues are addressed in Section 2.4: backtracking and initialization of the algorithm. Specifically, the section discusses how backtracking assures global convergence by providing a mechanism to recover from incorrect moves, and how the initial search can be sped up by incorporating domain knowledge and application-specific heuristics. The final section discussing implementation of the NP method is Section 2.5, which presents alternatives for specifying a promising index. Finally, Section 2.6 proves finite time convergence of the NP method and analyzes the behavior of the algorithm. This section can be skipped for those readers primarily interested in applications.

Chapter 3 focuses on the special case where the objective function is not known analytically but must be estimated and is hence noisy. This introduces additional challenges but the NP method can still be applied effectively. This chapter discusses what makes the NP method effective for problems with noisy performance, provides a convergence analysis, and suggests how the NP method is best implemented for such problems. This chapter can be skipped unless the reader is interested in problems with such noisy performance since only Chapter 7, Chapter 9, and Chapter 12 will make use of the results presented in this chapter.

Mathematical programming methods have been shown to effectively solve numerous large-scale problems, and in Chapter 4 we show how such methods can be effectively incorporated into the NP framework to improve its efficiency.

In this chapter we start by exploring the connections between the NP method and two traditional mathematical programming methods: branch-and-bound and dynamic programming. We then show that even for problems that are too large or complex for exact methods to be efficient, it is often possible to incorporate such methods into the NP framework by using mathematical programming to solve a relaxed or partial problem and then incorporate the solution into the NP method to improve its efficiency. This can be done by using the solution to either define a more effective partitioning or to help generate better feasible solutions. Thus, the NP method and mathematical programming are found to be highly complementary.

Similar to mathematical programming in Chapter 4, in Chapter 5 we demonstrate how various random search methods, metaheuristics, and local search can be incorporated into the NP framework. To illustrate how to incorporate other metaheuristics into the NP framework, we present three well-known metaheuristics: genetic algorithms, tabu search, and ant colony optimization. Each of these can be thought of as an improving search heuristic in that it constructs a sequence of feasible solutions, although each has a mechanism that allows an escape from a local optimum so that the sequence is not necessarily always improving. Hence, it is natural to incorporate these heuristics into the NP framework by using them to generate high-quality feasible solutions from each region that is being considered. Any local search heuristic can in a similar way be used to generate feasible solutions within the NP framework. But we will see how local heuristics may also be used to define an improved partitioning and heuristics can therefore be used to improve the efficiency of the NP method through both partitioning and generating feasible solutions.

The application section of the book consists of seven independent chapters illustrating how the NP method can be used to solve problems that arise in a broad range of application areas. Leading off, Chapter 6 looks at a very complex production scheduling problem that arises where there are flexible resources that must be scheduled simultaneously to the jobs that are to be completed, namely the Parallel-Machine Flexible-Resource Scheduling (PM-FRS) problem. The chapter shows how the NP method can be implemented for the PMFRS problem and how the problem can be reformulated so that the NP method may take advantage of the special structure of the problem in both the partitioning and the generation of good feasible solutions. To that end, a new random sampling algorithm that biases the sampling towards good schedules and a simple resource allocation improvement heuristic are developed. The numerical results indicate that high-quality schedules may be obtained using the NP method and it is shown that it is particularly useful for large-scale problems.

With the proliferation of massive data gathering and storage, data mining for extracting meaningful information from the resulting databases has become increasingly important. Chapter 7 looks at a problem that arises frequently in practical data mining projects, namely the feature selection problem briefly introduced above. The chapter shows that the NP method can be effective in obtaining high-quality feature subsets with reasonable time and that by using intelligent partitioning the efficiency of the algorithm can be improved by an order of magnitude. Taking advantage of the fact that the NP method is effective even if the objective function is noisy, this chapter also develops an adaptive version of the NP algorithm that uses only a sample of instances in each iteration and is consequently capable of scaling to large databases. This is possible because of the backtracking aspect of the NP method that allows the algorithm to recover from incorrect moves made due to decisions being made based on a relatively small fraction of all instances. The numerical results indicate that the NP method requires using only a small fraction of instances in each step to obtain good solutions and this fraction tends to decrease as the problem size increases, making it very scalable to large-scale feature selection problems.

An important problem for many organizations today is the design of their supply chain network. Chapter 8 considers how to apply the NP method to difficult optimization problems that arise in this context. The computational results reported in the chapter demonstrate that the NP method is capable of efficiently producing very high-quality solutions to distribution system design problems. In particular, the NP method is very effective for large-scale problems, and for such problems it is demonstrated to be significantly faster and generates better feasible solutions than either general-purpose combinatorial optimizers (such the branch-and-cut solver within CPLEX) or specialized approaches such as those based on Lagrangian relaxation. The results reported in this chapter also illustrate that the NP framework can effectively combine problem-specific heuristics with mixed integer programming (MIP) tools.

Many important problems that arise in health care delivery have recently been increasingly addressed using analytical techniques, and this includes the planning of radiation treatments. One of the problems that arises when planning such treatments is the beam angle selection (BAS) problem briefly introduced above, and Chapter 9 treats this problem in more detail. In this chapter we demonstrate that the NP method provides an effective framework for obtaining high-quality solutions to the BAS problem. Furthermore, relative to good quality beam angle sets constructed via expert clinical judgement and other approaches the beam sets generated via NP showed significant improvement in performance as measured by reduction in radiation delivered to non-cancerous organs-at-risk near the tumors. Thus, in addition to providing a method for automating beam angle selection, the NP method yields higher quality beam sets that significantly reduce radiation damage to critical organs.

Chapter 10 deals with a problem in another important area, namely transportation and logistics. In particular, the chapter provides a mixed integer programming (MIP) formulation of the local pickup and delivery problem (LPDP) and shows how a hybrid NP that utilizes the lower bounds of the MP can solve very difficult instances of this problem. This implementation illustrates how even when mathematical programming techniques cannot effectively solve a given problem due to its size and complexity, they can be incorporated into the NP framework through both partitioning and a lower bound biased sampling approach to significantly improve the efficiency of the NP method. The numerical results reported in this chapter demonstrate that the hybrid NP algorithm is more effective than a standard mathematical programming approach, in particular for large-scale problems.

Chapter 11 deals with a complex extension of the classic job shop scheduling problem, where bill-of-material and work-shift constraints are also accounted for in the the formulation. This problem is motivated by observations of real job shop systems, and this chapter illustrates how the NP method can effectively handle realistic problems with very complex constraints. The NP algorithm developed for this problem utilizes intelligent partitioning to impose structure on the search space, and uses an innovative sampling strategy to generate high-quality solutions subject to complex constraints. Numerical results using real industry data are reported.

The final chapter considers problems where uncertainty plays a key role. Specifically, the design of discrete event systems gives rise to many resource allocation problems, and in Chapter 12 we discuss two such examples, namely buffer allocation in communication networks and resource allocation in manufacturing systems. In both cases the objective function is stochastic. We show how metaheuristics such as tabu search can be integrated into the NP framework and how the hybrid method can be used to effectively deal with such uncertainty.

# Part I

Methodology

# 2

## The Nested Partitions Method

This chapter lays the groundwork for subsequent chapters as we introduce basic characteristics of the *nested partitions* (NP) optimization framework for solving large-scale optimization problems. This method systematically *partitions* the feasible region into subregions and moves from one region to another based on information obtained by randomly generating feasible sample solutions from each of the current regions. The method keeps track of which part of the feasible region is the most promising in each iteration and the number of feasible solutions generated, and thus the computational effort is always concentrated in this most promising region. The NP method is therefore particularly efficient for problems where the feasible region can be partitioned such that good solutions tend to be clustered together and the corresponding regions are thus natural candidates for concentrating the computation effort. In this chapter we discuss how to effectively partition to achieve such a structure, how to randomly generate feasible solutions, how to recover from incorrect moves through backtracking, and show that for any combinatorial optimization problem the NP method finds the optimal solution in finite time.

## 2.1 Nested Partitions Framework

Consider a combinatorial optimization problem (COP) or a mixed integer program (MIP) where there may exist many locally optimal solutions. The set of feasible solutions, called the feasible region, is denoted $X$ and a linear or nonlinear objective function $f : X \rightarrow \mathbf{R}$ is defined on this set. In mathematical notation we are interested in finding a feasible solution $x^* \in X$ that globally minimizes the objective function, that is, solving the following problem:

$$\min_{x \in X} f(x). \tag{2.1}$$

The value of the objective function is denoted $f^*$, so

$$f^* = f(x^*) \leq f(x), \forall x \in X.$$

When (2.1) is a COP then $X$ is finite and the problem is mathematically trivial in the sense that all we need to do is to enumerate all the solutions in $X$ and determine which has the best performance value. In practice, this is not possible due to the large number of feasible solutions. Some problems have a special structure that can be exploited to find the optimal solution without checking all the alternatives. For example, when $f$ is linear and $X$ is defined by linear constraints, that is, $X = \{x \in Z_+^n : Ax \le b\}$ then (2.1) is an integer program (IP) and exact optimization methods can often be used to solve at least small to moderately large problems. However, many real problems are either not sufficiently structured or too large for this to be possible. Such complex large-scale optimization problems are the subject of this book.

In each iteration of the NP algorithm we assume that there is a region (subset) of $X$ that is considered *the most promising*. We partition this most promising region into some fixed number of $M$ subregions and aggregate the entire *complimentary region* into one region, that is, all the feasible solutions that are not in the most promising region. At each iteration, we therefore consider $M + 1$ subsets that are a partition of the feasible region $X$, namely they are disjoint and their union is equal to $X$. Each of these $M + 1$ regions is sampled using some random sampling scheme to generate feasible solutions that belong to that region. The performance function values of the randomly generated samples are used to calculate the *promising index* for each region. This index determines which region is the most promising region in the next iteration. If one of the subregions is found to be best, this region becomes the most promising region. If the complimentary region is found to be best the region that was the most promising region in the previous iteration becomes the most promising region again, that is, the algorithm *backtracks* to a previous solution. The new most promising region is then partitioned and sampled in the same fashion.

Unless there is prior domain knowledge that can be utilized, the algorithm initializes by assuming that all parts of the feasible region are equally promising, that is, the entire feasible region $X$ is the most promising region. Since the complementary region is empty, it is sufficient to sample from the $M$ subregions in the first iteration, or in any iteration where $X$ is considered the most promising region. It is also clear that when $X$ is finite eventually there will be regions that contain only a single solution. We call such singleton regions regions of *maximum depth*, and more generally, talk about the *depth* of any region. This is defined iteratively in the obvious manner, with $X$ having depth 0, the depth of the subregions of $X$ being one, and so forth. Then the problem is infinite, e.g. when solving a MIP, we define the maximum depth to correspond to the smallest desired sets.

Assume that a method for partitioning has been fixed. This means that the number of subregions has been decided, as has what rule is followed in partitioning any given region into subsets. If necessary, a maximum depth $d^*$ has been specified. We call a subset that is constructed using this fixed partitioning method a *valid region*. If a valid region $\sigma$ is formed by partitioning a

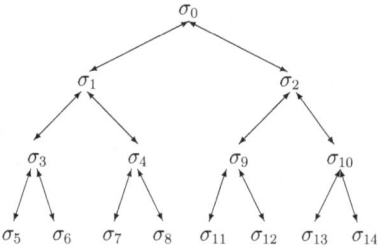

**Fig. 2.1.** Example of partitioning for the NP method.

valid region $\eta$, then $\sigma$ is called a *subregion* of region $\eta$. The following examples illustrate how partitioning can be done for some simple problems.

**Example 2.1.** Consider a feasible region that consists of 8 points $\sigma_0 = X = \{1, 2, 3, 4, 5, 6, 7, 8\}$ and in each iteration partition the current most promising region into two disjoint sets (see Figure 2.1). At the first iteration, $\sigma_1 = \{1, 2, 3, 4\}$ and $\sigma_2 = \{5, 6, 7, 8\}$ are sampled. Assume that the promising index (the sampling result) in $\sigma_1$ is better than in $\sigma_2$; select $\sigma_1$ as the most promising region and further partition $\sigma_1$ to obtain $\sigma_3 = \{1, 2\}$ and $\sigma_4 = \{3, 4\}$. At the second iteration, $\sigma_3$, $\sigma_4$, and their complimentary region, $\sigma_2$, are sampled. If the promising index of $\sigma_3$ (or $\sigma_4$) is the best, then select $\sigma_3$ to be the most promising region and partition $\sigma_3$ further into another two subregions $\sigma_5 = \{1\}$ and $\sigma_6 = \{2\}$ (or select $\sigma_4$ to be the most promising region and partition $\sigma_4$ into another two subregions $\sigma_7 = \{3\}$ and $\sigma_8 = \{4\}$). If the promising index of $\sigma_2$ is the best, then select $\sigma_0$ as the most promising region. Now assume that $\sigma_3$ is the most promising region, at the third iteration, $\sigma_5$, $\sigma_6$, and their complimentary region $(\sigma_0 \setminus (\sigma_5 \cup \sigma_6))$ are sampled. If the promising index of $\sigma_5$ (or $\sigma_6$) is the best, then select $\sigma_5$ (or $\sigma_6$) as the most promising region. If the promising index of the complimentary region is the best, then select $\sigma_1$ as the most promising region. As the algorithm evolves, a sequence of most promising regions $\{\sigma(k)\}$ will be generated. Here $\sigma(k)$ is the most promising region in the $k$th iteration.

**Example 2.2.** Assume the same feasible region and partitioning as in Example 2.1 (see Figure 2.1). The following sequences are two of the possible sequences of the most promising regions:

$$\sigma_0 \to \sigma_1 \to \sigma_4 \to \sigma_1 \to \sigma_3 \to \sigma_6,$$

$$\sigma_0 \to \sigma_1 \to \sigma_0 \to \sigma_2 \to \sigma_9 \to \sigma_{12}.$$

Now consider this sequence,

$$\sigma_0 \to \sigma_1 \to \sigma_4 \to \sigma_3.$$

This is not a possible sequence since the algorithm cannot move directly from $\sigma_4$ to $\sigma_3$.

As can be seen from these examples, one of the key elements of the NP method is to shift the focus from specific points in the feasible region $X$ to a space of subsets of $X$, namely the space of all valid regions. We denote this space $\Sigma$. Therefore, in addition to the objective function that is defined on $X$, we need to have a set performance function defined on $\Sigma$, the valid subregions of $X$. We can then use this set function to select the most promising region. We note that this shift of focus is similar to that employed by any branching procedure, such as branch-and-bound or branch-and-cut. However, such branching algorithms focus the computational effort on obtaining a lower bound for each subset, whereas the NP method defines the promising index in terms of feasible solutions that are generated using a random sampling procedure.

To complete the notation, we let $\sigma(k)$ denote the most promising region in the $k$th iteration, and let $d(k)$ denote the depth of $\sigma(k)$. With all the necessary notation in hand, the algorithm is described below. The special cases of being at minimum or maximum depth are considered separately, but first the general case of iteration $k$ is discussed where $0 < d(k) < d^*$ and $d^*$ is the maximum depth. We refer to this algorithm as the *Pure NP Algorithm* to distinguish it from hybrid algorithms that will be introduced later in this book.

**Algorithm** *Pure NP*

1. **Partitioning.** Partition the most promising region $\sigma(k)$ into $M$ subregions $\sigma_1(k), ..., \sigma_M(k)$, and aggregate the complimentary region $X \setminus \sigma(k)$ into one region $\sigma_{M+1}(k)$.

2. **Random sampling.** Randomly generate $N_j$ sample solutions from each of the regions $\sigma_j(k)$, $j = 1, 2, ..., M + 1$:

$$x_1^j, x_2^j, ..., x_{N_j}^j, \quad j = 1, 2, ..., M + 1.$$

Calculate the corresponding performance values:

$$f(x_1^j), f(x_2^j), ..., f(x_{N_j}^j), \quad j = 1, 2, ..., M + 1.$$

3. **Calculate promising index.** For each region $\sigma_j$, $j = 1, 2, ..., M + 1$, calculate the *promising index* as the best performance value within the region:

$$I(\sigma_j) = \min_{i=1,2,...,N_j} f(x_i^j), \quad j = 1, 2, ..., M + 1. \tag{2.2}$$

4. **Move.** Calculate the index of the region with the best performance value.

$$\hat{j}_k \in \arg\min_{j=1,...,M+1} I(\sigma_j), \quad j = 1, 2, ..., M + 1. \tag{2.3}$$

If more than one region is equally promising, the tie can be broken arbitrarily. If this index corresponds to a region that is a subregion of $\sigma(k)$, that is $\hat{j}_k \leq M$, then let this be the most promising region in the next iteration

$$\sigma(k+1) = \sigma_{\hat{j}_k}(k) \tag{2.4}$$

Otherwise, if the index corresponds to the complimentary region, that is $\hat{j}_k = M + 1$, backtrack to the previous most promising region:

$$\sigma(k+1) = \sigma(k-1). \tag{2.5}$$

For the special case of $d(k) = 0$ (that is, $\sigma(k) = X$), the steps are identical except that there is no complimentary region. The algorithm thus generates feasible sample solutions from the subregions and in the next iteration moves to the subregion with the best promising index. For the special case of $d(k) = d^*$ there are no subregions. The algorithm therefore generates feasible sample solutions from the complimentary region and either backtracks or stays in the current most promising region.

It is apparent that this basic implementation of the NP method is very simple. It is indeed this simplicity that gives it the flexibility to effectively incorporate application-specific structure and methods while providing a framework that guides the search and enables meaningful convergence analysis. In the next four sections we discuss the implementation of each step in more detail.

## 2.2 Partitioning

The partitioning is of paramount importance to the efficiency of the NP method because the selected partition imposes a structure on the feasible region. When the partitioning is done in such a way that good solutions are clustered together, then those subsets tend to be selected by the algorithm with relatively little effort. On the other end of the spectrum, if the optimal solution is surrounded by solutions of poor quality it is unlikely that the algorithm will move quickly towards those subsets. For some problems a simple partition may automatically achieve clustering of good solutions but for most practical applications more effort is needed. We will see how it is possible to partition effectively by focusing on the most difficult decisions and how both heuristics and mathematical programming can be applied to find partitions that improve the efficiency of the algorithm. We refer to such partitions as *intelligent partitioning* in order to distinguish it from *generic partitioning* that partitions the feasible region without considering domain knowledge, the objective function, or other special structure.

### 2.2.1 A Generic Partitioning Method

We illustrate a generic partitioning through the traveling salesman problem (TSP). This classic COP can be described as follows: Imagine a traveling salesman who must visit a set of cities. The objective is to minimize the distance traveled, while the constraints assure that each city is visited exactly once (assignment constraints) and that the selected sequence of cities forms a connected tour (subset elimination constraints). Without the subset elimination constraints the TSP reduces to a simple assignment problem, whereas with these constraints it is a NP-hard problem, which implies that it is unlikely that a polynomial time algorithm exists for its solution.

Assume that there are $n + 1$ cities. For a generic partitioning method, arbitrarily choose city 0 as the starting point and label the other cities as $1, 2, 3, ...., n$. The feasible region becomes all permutations of $\{1, 2, 3, ...., n-1\}$,

$$X = \left\{ x \in Z_+^n : 1 \leq x_i \leq n, x_i \neq x_j \text{ if } i \neq j \right\}.$$

First, partition the feasible region into $n$ regions by fixing the first city on the tour to be one of $1, 2, ..., n$. Partition each such subregion further into $n - 1$ regions by fixing the second city as any of the remaining $n - 1$ cities on the tour. This procedure can be repeated until all the cities on the tour are fixed and the maximum depth is reached. In this way the subregions at maximum depth contain only a single solution (tour). Figure 2.2(a) illustrates this approach.

Clearly there are many such partitions. For example, when choosing city 0 as the starting point, instead of fixing the first city on the tour, fix any $i$th city on the tour to be one of cities $1, 2, ..., n$ (see Figure 2.2(b)). This partition provides a completely different set of subregions, that is, the set $\Sigma$ of valid regions will be different than before.

This is a generic partition because it does not take advantage of any special structure of the TSP and also does not take the objective function into account. It simply partitions the feasible region without considering the performance of the solutions in each region of the partition. It is intuitively appealing that a more efficient implementation of the NP method could be achieved if the objective function was considered in the partitioning to assure that good solutions are clustered together.

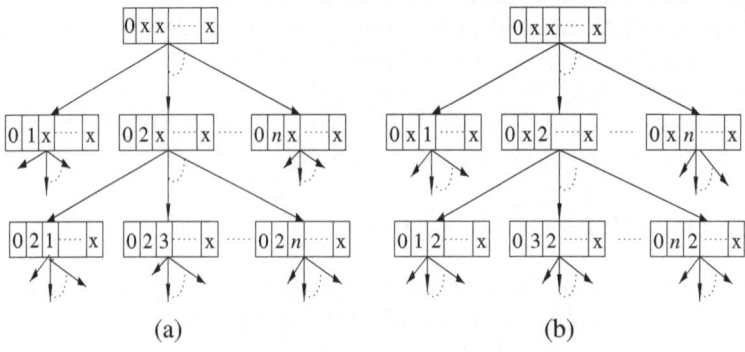

(a)                              (b)

**Fig. 2.2.** Two generic partitions.

### 2.2.2 Intelligent Partitioning for TSP

The generic partitioning does not consider the objective function when partitioning the feasible region. This may lead to difficulties in distinguishing between regions and consequently the algorithm may not efficiently locate where to concentrate the computational effort. If the NP method is applied using the above partitioning, it may backtrack frequently and not settle down in a particular region. On the other hand, the NP method is likely to perform more efficiently if good solutions tend to be clustered together for a given partitioning. To impose such structure, consider the following partitioning scheme through a simple example.

**Example 2.3.** Assume $n = 5$ cities are defined by the undirected graph in Figure 2.3. As an initialization procedure store the edges in an adjacency list and sort each of the linked lists that are connected to the cities (see the following table). For example, in the following adjacency list, the first row provides a linked list for city $A$, that is $E$ is the city closest to $A$, $C$ is the city second closest to $A$, $B$ is the city next closest to $A$, and $D$ is the city farthest from $A$.

| City | Closest two | | Next two | |
|---|---|---|---|---|
| $A \rightarrow$ | $E \rightarrow$ | $C \rightarrow$ | $B \rightarrow$ | $D$ |
| $B \rightarrow$ | $C \rightarrow$ | $A \rightarrow$ | $D \rightarrow$ | $E$ |
| $C \rightarrow$ | $A \rightarrow$ | $B \rightarrow$ | $D \rightarrow$ | $E$ |
| $D \rightarrow$ | $C \rightarrow$ | $E \rightarrow$ | $A \rightarrow$ | $B$ |
| $E \rightarrow$ | $A \rightarrow$ | $C \rightarrow$ | $B \rightarrow$ | $D$ |

This adjacency list becomes the basis of the intelligent partitioning. The entire region is all paths that start with the city A (chosen arbitrarily). If in each iteration the solution space is partitioned into $M = 2$ subregions then

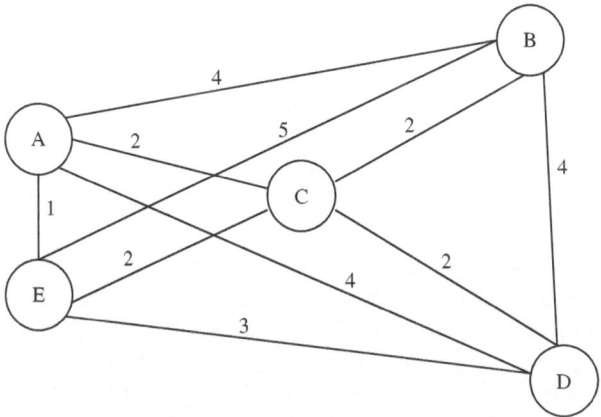

**Fig. 2.3.** An example TSP problem.

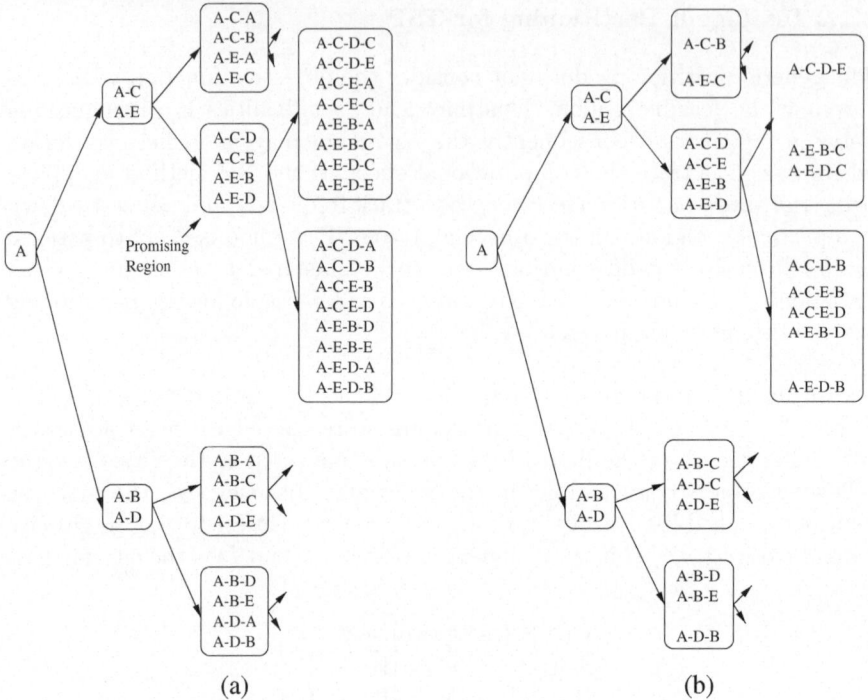

**Fig. 2.4.** Intelligent partitioning for the TSP.

the first subregion consists of all the paths that start with either (A,E) or (A,C) as the first edge. The second subregion consists of all the paths that start with (A,B) or (A,D) as the first edge.

Now assume that the first subregion is chosen as the most promising region (see Figure 2.4). Then the first subregion of that region is the one consisting of all paths that start with (A,E,A),(A,E,C),(A,C,A) or (A,C,B). The second region can be read from the adjacency list in a similar manner. Notice that one of these conditions creates an infeasible solution so there is no guarantee that all paths in a subregion will be feasible. It is, however, easy to check for feasibility during the sampling stage, and in fact this must always be done.

### 2.2.3 Intelligent Partitioning for Feature Selection

The previous section describes intelligent partitioning for the TSP that uses the objective function directly. We now illustrate a different approach for the feature selection problem introduced in Chapter 1.

Let $A = \{a_1, a_2, ..., a_n\}$ denote the set of all features (variables). Recall that in a data mining project, the feature selection problem determines which elements of $A$ are selected to be used by a subsequent inductive learning algorithm. In other words, the decision variables are

$$x_i = \begin{cases} 1 \text{ if the } i\text{th feature } a_i \in A \text{ is included,} \\ 0 \text{ otherwise.} \end{cases}$$

Thus, given a current set $\sigma(k)$ of potential feature subsets, partition the set into two disjoint subsets

$$\sigma_1(k) = \{A \in \sigma(k) : a \in A\}, \qquad (2.6)$$

$$\sigma_2(k) = \{A \in \sigma(k) : a \notin A\}. \qquad (2.7)$$

Hence, a partition is defined by a sequence of features $a_1, a_2, \ldots, a_n$, which determines the order in which the features are either included or excluded (see Figure 2.5). According to the goals of a good partition, the order of the features should be selected such that the features that best separate good feature subsets from poor sets are selected first. In other words, if there is a feature that must be included in any high-quality feature subset, or, vice versa, a feature that should not be included, it is advantageous to select this feature early. This calls for a reordering of the features in order to impose the best possible structure. There are a number of strategies that have been developed to measure the importance of features in classification (Shih 1999), including the information gain that is obtained by knowing the value of each of the features, and this is the method that we utilize here.

Recall that the eventual goal of the feature selection is to determine a set of features in a training data set $T$ such that when an inductive learning algorithm is applied to this training set, a high quality model results. Now suppose a training set $T$ of $m$ instances contains $s_{ij}(a)$ instances where feature $a$ is set to its $j$th value and the instance is classified as the $i$th class. The total number of instances where $a$ is set to the $j$th value is then $S_j(a) = \sum_{i=1}^{c} s_{ij}(a)$,

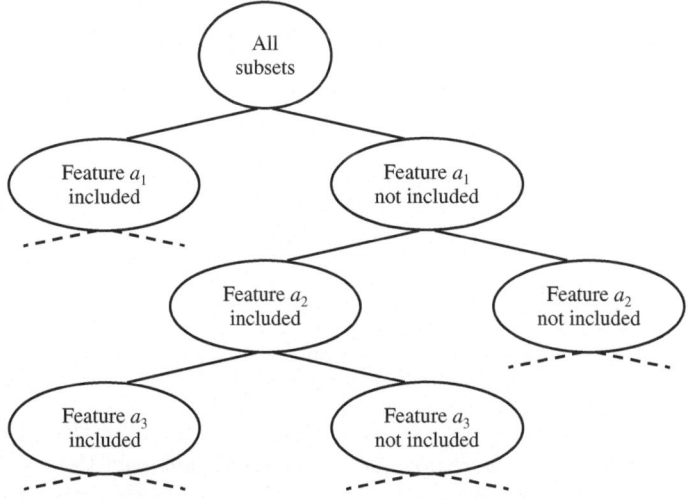

**Fig. 2.5.** Partitioning for the feature selection problem.

where $c$ is the total number of classes. For this training set, the expected information that is needed to classify a given instance is given by

$$I(T) = -\sum_{i=1}^{c} p_i \log_2(p_i),$$

(2.8)

where $p_i = \dfrac{\sum_j s_{ij}(a)}{m}$ is the fraction of instances that belong to the $i$th class. The information gain of a feature is the expected amount by which (2.8) is reduced if the value of the feature is known. It is calculated based on the entropy of the feature

$$E(a) = \sum_{j=1}^{v} q_j(a) I_j(a)$$

(2.9)

where $v$ is the number of distinct values that feature $a$ can take, and $q_j = \dfrac{S_j(a)}{m}$, the relative frequency of the $j$th value in the training set, is the weight when $a$ is set to its $j$th value

$$I_j(a) = -\sum_{i=1}^{c} p_{ij} \log_2(p_{ij}),$$

(2.10)

where $p_{ij} = \dfrac{s_{ij}(a)}{S_j(a)}$ is the proportion of instances with $j$th value of feature $a$ that belong to the $i$th class. Then the information gain of feature $a$ is (Quinlan 1986)

$$Gain(T, a) = I(T) - E(a),$$

(2.11)

that is, the expected reduction in entropy that would occur if we knew the value of feature $a$. Note that the feature with the highest information gain has the lowest entropy value.

The maximum information gain, or equivalently the minimum entropy, determines a ranking of the features. Thus, we select

$$a_{[1]} = \arg\min_{a \in A} E(a),$$

$$a_{[2]} = \arg\min_{a \in A \setminus \{a_{[1]}\}} E(a),$$

$$\vdots$$

$$a_{[n]} = \arg\min_{a \in A \setminus \{a_{[1]}, \ldots, a_{[n-1]}\}} E(a).$$

This new feature order $a_{[1]}, a_{[2]}, \ldots, a_{[n]}$ defines a partition for the NP method that we call the *entropy partition* and is an intelligent partitioning for the feature selection problem. We note that we chose to consider entropy to define the partition due to past success of using this measure for feature selection.

However, any other method for evaluating the value of individual features could be used in a similar manner.

We will revisit this example in Chapter 7 and show how this intelligent partitioning can achieve an order of magnitude improvement in the efficiency of the NP method.

### 2.2.4 General Intelligent Partitioning

Intelligent partitioning methods will in general be application-dependent, but it may be possible to devise some general intelligent partitioning methods that perform well for a large class of problems. One such method would be based on generalizing the ideas from the last section.

To develop a general intelligent partitioning scheme, we use the diversity idea introduced above, which originates in information theory and is well known in areas such as machine learning and data mining. For this purpose a solution needs to be classified as being the same from the point of view of performance. A natural way to think about this is to say that two solutions are the same if there is little difference in their objective function values. Thus, a valid subregion where there are many solutions with significantly different objective function values is considered diverse, and vice versa. Diverse subregions are undesirable as they make it difficult to determine which subregion should be selected in the next move.

To use traditional diversity measures, classify each solution into one category. First specify a small value $\epsilon > 0$ such that two solutions $x^1, x^2 \in X$ can be defined as having similar performance if $|f(x^1) - f(x^2)| < \epsilon$. Then construct categories such that all solutions in each category are similar in this sense, and for any two categories there is at least one solution in each such that both are dissimilar.

The following scheme can now be used to construct an intelligent partitioning:

1. Use random sampling to generate a set of $M_0$ sample solutions.
2. Evaluate the performance $f(x)$ of each one of these sample solutions, and record the average standard error $\bar{s}^2$.
3. Construct $g(\bar{s}^2)$ intervals or categories for the sample solutions.
4. Let $S_l$ be the frequency of the $l$th category in the sample set, and $q_l = \frac{S_l}{M_0}$ be the relative frequency.
5. Let $i = 1$.
6. Fix $x_i = x_{ij}$, $j = 1, 2, ..., m(x_i)$.
7. Calculate the proportion $p_l$ of solutions that falls within each category, and use this to calculate the corresponding entropy value:

$$E(i) = \sum_{l=1}^{g(\bar{s}^2)} q_l(i) \cdot I_l(i) \tag{2.12}$$

where

$$I_l(i) = -\sum_{l=1}^{g(\bar{s}^2)} p_{ij} \log_2(p_{ij}), \qquad (2.13)$$

where $p_{ij}$ is the proportion of samples with $x_i = x_{ij}$.

8. If $i = n$, stop; otherwise let $i = i + 1$ and go back to step 6.

A high entropy value indicates high diversity, so it is desirable to partition by fixing the lowest entropy dimensions first. Thus, order the dimensions according to their entropy values

$$E(x_{[1]}) \leq E(x_{[2]}) \leq ... \leq E(x_{[n]}), \qquad (2.14)$$

and let this order determine the intelligent partition.

Note that we would apply this procedure before starting the actual NP method. Significant computational overhead may be incurred when determining an intelligent partition in this manner, but for difficult applications it is often worthwhile to expend such computational effort developing intelligent partitioning. This imposes useful structure that can thus improve the efficiency of the NP method itself, which often vastly outweighs the initial computational overhead.

## 2.3 Randomly Generating Feasible Solutions

In addition to using domain understanding to devise partitioning that imposes a structure on the feasible region, the other major factor in determining the efficiency of the NP method is the method employed for generating feasible solutions from each region. The Pure NP Algorithm prescribes that this should be done randomly but there is a great deal of flexibility both in how those random samples should be generated and in how many random samples should be obtained.

The goal should be for the algorithm to frequently make the correct move, that is, either to move to a subregion containing a global optimum or to backtrack if the current most promising region does not contain a global optimum. In the theoretical ideal, the correct move will always be made if the best feasible solution is generated in each region. This is of course not possible except for trivial problems, but in practice the chance of making the correct move can be enhanced by (i) biasing the sampling distribution so that good solutions are more likely to be selected, (ii) incorporating heuristic methods to seek out good solutions, and (iii) obtaining a sufficiently large sample. We will now explore each of these issues.

### 2.3.1 Biased Random Sampling

We illustrate some simple random sampling methods for generating feasible solutions to the TSP. Assume that the generic partitioning (Figure 2.2) is used and the current most promising region is of depth $k$. This means that the first $k$ edges in the TSP tour have been fixed. Generating a sample solution from this region entails determining the $n - k$ remaining edges. One approach would be simply to select the edges consecutively, such that each feasible edge has equal probability of being selected (uniform sampling). However, this approach may not give good results in practice. The reason is the same as for why a generic partitioning may be inefficient, that is, uniform sampling considers only the solution space itself.

To incorporate the objective function into the sampling, consider the following biased sampling schemes. At each iteration, weights $(w_{j_{i-1},j_l})$ are calculated and assigned to each of the remaining cities that need to be determined. The weight is inversely proportional to the cost of the edge from city $j_l$ to the city $j_{i-1}$. Specifically we can select the weights as

$$w_{j_{i-1},j_l} = \frac{\left(c_{j_{i-1},j_l}\right)^{-1}}{\sum_{h=i}^{n} \left(c_{j_{i-1},j_h}\right)^{-1}},$$

where the weights have been normalized so that the sum is one. The next edge can now be selected by uniformly generating a number $u$ between zero and one and comparing this number with the weights, that is, if

$$\sum_{m=i}^{i^*} w_{j_{i-1},j_m} \leq u < \sum_{m=i}^{i^*+1} w_{j_{i-1},j_m}$$

then city $i^*$ is selected and $(j_{i-1}, j_{i^*})$ becomes the next edge.

This is a randomized procedure for generating feasible sample solutions. Each edge has a positive probability of being selected next and thus for every region, all of the feasible tours in the region have a positive probability of being generated using this procedure. However, this probability is no longer uniform. The probability has been biased so that low-cost edges are selected with higher probability and tours with many low-cost edges are therefore generated with higher probability. In our computational experience such biased sampling approaches will tend to significantly improve the efficiency of the NP method.

It is important to note that sampling is very flexible when dealing with very hard constraints. Say for example that the TSP has some additional constraints such as time windows or restrictions on the order in which the cities must be visited. Such constraints can be very difficult to deal with using mathematical programming techniques, but only a minor modification of the sampling procedure would be needed to assure that only feasible solutions are generated. Thus, the use of sampling makes the NP method extremely effective when dealing with complex constraints. As noted in Section 1.3.1 of Chapter 1, many problems that arise in complex applications have both easy and complex constraints and using the NP method, it is possible to use sampling to deal with the complex constraints, while using exact methods to deal with the easy constraints.

### 2.3.2 Incorporating Heuristics in Generating Solutions

In addition to biasing the sampling distribution, it may be possible to quickly generate good feasible solutions by applying a (randomized) heuristic search. This can be done for example with a simple local search such as in the following algorithm that generates $N$ feasible solutions.

**Local Search Sampling**

1. Obtain one random sample using either uniform or weighted sampling.
2. Obtain $N_j - 1$ more samples by making small perturbations of the sample generated in the first step.

To illustrate this approach consider how the local search sampling can be applied to the TSP. The second step could for example involve randomly selecting two edges and connecting the first vertex of the first edge to the first vertex of the second edge, and connecting the second vertex of the first edge to the second vertex of the second edge. This technique is similar to a 2-opt exchange but does not consider whether the performance is improved by this exchange. Other more complicated variants, with more than two edges selected at random, are easily obtained in a similar fashion. Step 2 is further illustrated in Figure 2.6. A sample obtained in step 1 is shown with the combination of solid and dotted lines. Then in step 2, select the edges (C,E) and (D,A) at random and replace them with the edges (C,D) and (E,A). The new edges are shown as dashed lines in Figure 2.6. Clearly this procedure provides us with a new sample solution with relatively little effort.

Instead of using an application-specific heuristic such as the one above, any other heuristic could be incorporated in the same manner. For example, we can incorporate the genetic algorithm (GA) cross-over operator into the sampling as follows for the TSP problem above. Start by randomly generating two sample solutions, say, $(A, E, C, D, B)$ and $(E, A, C, B, D)$, called the parents. Select a cross-over point, say after the second city on the tour, and generate two new solutions called the children, namely $(A, E, C, B, D)$, where the first two elements come from the first parent and the other three from the second, and $(E, A, C, D, B)$, where the first two elements come from the second parent

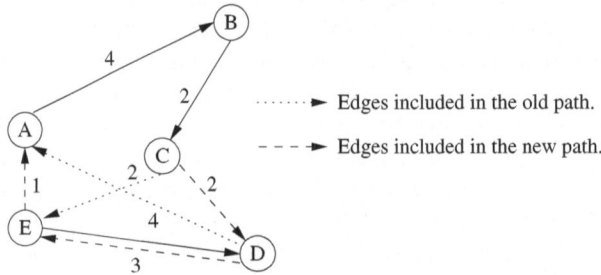

**Fig. 2.6.** The second step in the two-step sampling scheme.

and the other three from the first. Of course the cross-over operator can be applied more than once and the other main GA operator of mutation can similarly be incorporated into the NP framework. Thus, an entire GA search can easily be used to generate high-quality feasible solutions from each region and the same is true for any other heuristic that is thought to perform well for a particular application.

We refer to implementations of the NP method that incorporate other general-purpose heuristics, such as a genetic algorithm, as *hybrid NP methods*. Such hybrids, including a detailed NP/GA hybrid will be explored further in Chapter 5 and numerous specific application examples are also presented in Part II of the book.

### 2.3.3 Determining the Total Sampling Effort

As might be expected, incorporating special structure and heuristics to generate feasible solutions results in finding better solutions. This in turn leads to the NP algorithm selecting the correct move more frequently and thus improves the efficiency of the search. The question still remains as to how many sample solutions are needed to make the correct choice with a sufficiently large probability.

In Chapter 3 we will show that it is possible to connect the minimum required probability of making a correct move to the desired probability that the optimal solution is eventually found. The number of feasible sample solutions required to assure this minimum probability depends on the variance of the performance of the generated solutions. In the extreme case, if the procedure that is used to generate feasible solutions always results in a solution that has the same performance then there is no advantage to generating more than one solution. Vice versa, if the procedure leads to solutions that have greatly variable performance then it may be necessary to generate many solutions to obtain a sufficiently good estimate of the overall performance of the region.

This observation motivates the following two-stage sampling approach. In the first stage, generate a small number of feasible solutions using uniform sampling, weighted sampling, local search sampling, or any other appropriate method for generating sample solutions. Calculate the variance of the performance of these solutions and then apply statistical selection techniques to determine how many total samples are needed to achieve the desired results. This procedure will be made more precise in Chapter 3 where we address this issue in detail (see Section 3.2).

## 2.4 Backtracking and Initialization

Another critical aspect of the NP method is the global perspective it achieves by generating solutions from the complimentary region and backtracking if necessary. Specifically, if the best feasible solution is found in the complimentary region, this is an indication that the incorrect move was made when $\sigma(k)$

was selected as the most promising region so the NP algorithm backtracks by setting $\sigma(k+1) = \sigma(k-1)$.

Backtracking is usually very easy to implement since some type of truncation is usually sufficient and we do not need to keep track of the previous most promising region. Say for example in a five-city TSP problem that the current most promising region is defined by the sequence of cities $B \leftarrow D \leftarrow C$ with the remaining cities undecided. Thus, the current most promising region can be written as

$$\sigma(k) = \{(B, D, C, x_4, x_5) : x_4, x_5 \in \{A, E\}, x_4 \neq x_5\}.$$

If backtracking is indicated, then the next most promising region becomes

$$\sigma(k+1) = \{(B, D, x_3, x_4, x_5) : x_i \in \{A, C, E\}, x_i \neq x_j \text{ if } i \neq j\}.$$

Thus, backtracking is simply achieved by truncating the sequence that defines the current most promising region. Similar methods can be used for most other problems, making backtracking possible with very little or no overhead.

Unless otherwise noted we will always assume that backtracking is done as above, that is, $\sigma(k+1) = \sigma(k-1)$. However, it is clear that it is possible to backtrack in larger step, for example by truncating two or three cities from the sequence that defines the current most promising region in the TSP application. The advantage is that this would enable the algorithm to reverse a sequence of incorrect moves more easily. On other hand it reverses several NP moves based on the results from one iteration and if the backtracking turns out to be incorrect, it would take several moves to get back to the previous point. For this reason we do not advocate this approach but rather that the focus be directed towards making each move correctly with high probability. As noted above, this can be done by developing intelligent partitioning and using biased sampling and heuristics to generate high-quality feasible solutions. In other words, by incorporating domain knowledge and special structure, incorrect moves become infrequent and it is thus always sufficient to simply backtrack one step to the previous most promising region.

As we will see in the next section it is backtracking that assures that the NP method converges to the globally optimal solution and does not become stuck at a local optimum. However, in practice excessive backtracking indicates an inefficient implementation of the NP method. If backtracking is correctly called for this implies that at least one incorrect move was previously made. Thus, by monitoring the amount of backtracking, it is possible to design adaptive NP algorithms. If excessive backtracking is observed, this indicates that more effort is needed to evaluate regions before a choice is made. More or higher-quality feasible solutions should thus be generated before a choice is made, which can be achieved with uniform random sampling, local search sampling, or any other appropriate method. In Chapter 7 we will develop such an adaptive NP algorithm in detail.

Finally, we note that although we will generally assume that the initial state of the search is to let the entire feasible region be the most promising, that is, $\sigma(0) = X$, this need not be the case. For example, if time is very limited and it is important to generate good solutions very quickly (and then possibly continue to generate better solutions), it may be worthwhile to initialize the search and set $\sigma(0) = \eta$, where $\eta \in \Sigma \setminus \{X\}$ is a partial solution determined using a heuristic or domain knowledge.

## 2.5 Promising Index

The final aspect of the NP method is the promising index that is used to select the next most promising region. This promising index should be based on the sample information obtained by generating feasible sample solutions from each region, but other information could also be incorporated.

Unless otherwise noted we will assume that the promising index for a valid region $\sigma \in \Sigma$ is based only on the set of feasible solutions $D_\sigma$ that are generated from this region, and as in (2.2) it is taken to be

$$I(\sigma) = \min_{x \in D_\sigma} f(x). \tag{2.15}$$

Other promising indices may also be useful. For example, instead of basing the promising index only on an extreme point, the promising index could be defined as the average sample performance

$$I(\sigma) = \frac{1}{|D_\sigma|} \sum_{x \in D_\sigma} f(x). \tag{2.16}$$

It may also be helpful to incorporate more advanced mechanisms, such as mathematical programming bounds or the output of a metamodel, into the promising index. Say for example that it is easy to solve a relaxation of the problem (2.1) using mathematical programming methods and hence obtain a lower bound $\underline{f}(\sigma)$ on the objective function. This lower bound can be combined with the upper bound $\min_{x \in D_\sigma} f(x)$ into a single promising index

$$I(\sigma) = \alpha_1 \cdot \underline{f}(\sigma) + \alpha_2 \cdot \min_{x \in D_\sigma} f(x), \tag{2.17}$$

where $\alpha_1, \alpha_2 \in \mathbf{R}$ are the weights given to the lower bound and upper bound, respectively. The lower bound could be obtained using any appropriate standard mathematical programming method, such as linear programming (LP) relaxation, Lagrangian relaxation, or a application-specific COP relaxation, such as relaxing the subset elimination constraints for a TSP.

However, for large-scale complex discrete problems where the NP method is the most useful, it is often not possible to obtain useful lower bounds. In such cases, a probabilistic bound may be useful, which may be obtained as follows: The process by which feasible solutions are generated from each region attempts to estimate the extreme point of the region. This may or may not include a heuristic search. Let $x^*(\sigma)$ denote the true extreme point (minimum) of a valid region $\sigma \in \Sigma$. Then the extreme performance $\hat{f}^*(\sigma) = f(x^*(\sigma))$ is estimated as

$$\hat{f}^*(\sigma) = \min_{x \in D_\sigma} f(x), \tag{2.18}$$

where the set of feasible solutions $D_\sigma$ could be based on pure random sampling, applying a local search heuristic to an initial random sample, or applying a population-based heuristic such as a genetic algorithm to a random initial population. Now assume that we generate several such sets $D_\sigma^1, ..., D_\sigma^n$ and calculate the corresponding extreme value estimates $\hat{f}_1^*(\sigma), ..., \hat{f}_n^*(\sigma)$. It is then possible to construct an overall estimate

$$\hat{f}_{\min}^* = \min_i \min_{x \in D_\sigma^i} f(x) \tag{2.19}$$

and a $1 - \alpha$ confidence interval $[l(D_\sigma^1, ..., D_\sigma^n), u(D_\sigma^1, ..., D_\sigma^n)]$ for the extreme value, that is,

$$P\left[f^*(\sigma) \in [l(D_\sigma^1, ..., D_\sigma^n), u(D_\sigma^1, ..., D_\sigma^n)]\right] = 1 - \alpha.$$

The left end of the confidence interval may thus be viewed as a *probabilistic lower bound* for the performance (extreme point) of the region

$$\underline{\hat{f}}(\sigma) = l(D_\sigma^1, ..., D_\sigma^n). \tag{2.20}$$

This can then be incorporated into the promising index similar to the exact bound above, namely,

$$I(\sigma) = \alpha_1 \cdot \underline{\hat{f}}(\sigma) + \alpha_2 \cdot \hat{f}_{\min}^*. \tag{2.21}$$

The estimation of the confidence interval is typically based on the assumption that the extreme values follow a Weibull distribution.

In addition to the exact or probabilistic lower bounds, many other considerations could be incorporated into the promising index. For example cost penalties obtained from other regions or the variability of performance within the region could be added. The latter would be of particular interest if we want to obtain not only good solutions but also robust solutions, that is, solutions that are such that small changes in the solution will not greatly change the performance. In general, any domain knowledge or application-appropriate technique can in a similar manner be incorporated into the promising index and used to guide the search more efficiently.

Finally, if rather than only considering the best sample point, we wish to take advantage of all of the sample points, as in the average promising index in equation (2.16), those can often be incorporated into a metamodel, such as a response surface or a neural network, that predicts the optimal solution based on the sample points. Thus, the promising index becomes:

$$I(\sigma) = \min_{x \in \sigma} M_{D_\sigma}(x), \tag{2.22}$$

where $M_{D_\sigma}$ is a metamodel trained on the sample points $D_\sigma$, for example a neural network.

# 2.6 Convergence Analysis

In this section we analyze the convergence of the NP algorithm. The first main convergence result is that the NP algorithm converges to an optimal solution of any COP in finite time. The proof is based on a Markov chain analysis that utilizes the fact that the sequence of most promising regions is an absorbing Markov chain and the set of optimal solutions corresponds exactly to the absorbing states. Since the state space of the Markov chain is finite it is absorbed in finite time. With some additional assumptions, this result can be generalized to apply to problems with infinite countable feasible regions (such as IPs) and even continuous feasible regions (such as MIPs).

The second main result is that the time until convergence can be bounded in terms of the size of the problem and the probability of making the correct move. This result will show that the expected number of iterations grows slowly in terms of the problem size, which explains why the NP algorithm is effective for large-scale optimization. On the other hand, as the probability of making the correct move decreases, the expected number of iterations increases exponentially. This underscores the need to increase this probability by incorporating special structure in both the partitioning and the method used for generating feasible solutions.

### 2.6.1 Finite Time Convergence for COPs

In this section we assume that the NP algorithm is applied to a combinatorial optimization problem (COP). We start by formally stating the Markov property.

**Proposition 2.1.** *Assume that the partitioning of the feasible region is fixed and $\Sigma$ is the set of all valid regions. The stochastic process $\{\sigma(k)\}_{k=1}^{\infty}$, defined by the most promising region in each iteration of the pure NP algorithm, is a homogeneous Markov chain with $\Sigma$ as state space.*

*Proof:* The partitioning defines the state space $\Sigma$ of all valid regions. The NP algorithm moves from one state (valid region) to another based on the solutions generated from each state. Since the method for generating solutions is randomized the sequence $\{\sigma(k)\}_{k=1}^{\infty}$ of valid regions is a stochastic process. Since it does not depend on the iteration the stochastic process is homogeneous. Finally, since only solutions generated in the current iteration are used to determine where to move next, $\{\sigma(k)\}_{k=1}^{\infty}$ has the memoryless property and is a homogeneous Markov chain.

We next show that the optimal solution(s) are absorbing states.

**Proposition 2.2.** *Assume that the partitioning of the feasible region is fixed and $\Sigma$ is the set of all valid regions. A state $\eta \in \Sigma$ is an absorbing state for the Markov chain $\{\sigma(k)\}_{k=1}^{\infty}$ if and only if $d(\eta) = d^*$ and $\eta = \{x^*\}$, where $x^*$*

*is a global minimizer of the original problem, that is, it solves equation (2.1) above.*

*Proof:* We will start with the "if" part. Assume $\zeta = \sigma(k) = \{\hat{x}\}$ where $f(\hat{x}) \leq f(x)$, $\forall x \in X$, then the transition probability of staying in $\zeta$ is given by

$$P_{\zeta\,\zeta} = P[\hat{f}_{min}(\zeta) \leq \hat{f}_{min}(X \setminus \zeta)]$$
$$= P[f(\hat{x}) \leq \hat{f}_{min}(X \setminus \zeta)] = 1.$$

The fact that the NP algorithm breaks ties at maximum depth by favoring the current region is used. This proves that $\zeta$ is an absorbing state. Now its converse is proved. It is clear that a state that is an absorbing state is always of maximum depth since the transition probability of staying in such state is zero. It can therefore be assumed that $\eta \in \Sigma$ such that $d(\eta) = d^*$. In this case $\eta$ has only one point, say $\eta = \{\tilde{x}\}$. If there exists a better point in the complimentary region, say $x^* \in X \setminus \eta$ such that $f(x^*) < f(\tilde{x})$, then one gets the following transition probabilities.

$$P_{\eta\,X \setminus \eta} = P[\hat{f}_{min}(X \setminus \eta) < \hat{f}_{min}(\eta)]$$
$$= P[\hat{f}_{min}(X \setminus \eta) < f(\tilde{x})]$$
$$\geq P[\text{Point } x^* \text{ is picked at random from the set } X \setminus \eta] > 0.$$

This shows that $\eta$ is not an absorbing state, and completes the proof.

The following theorem is now immediate.

**Theorem 2.3.** *The NP algorithm converges almost surely to a global minimum of the optimization problem given by equation (2.1) above. In mathematical notation, the following equation holds:*

$$\lim_{k \to \infty} \sigma(k) = \{x^*\} \quad a.s. \tag{2.23}$$

*where*

$$x^* \in \arg\min_{x \in \Theta} f(\theta).$$

*Proof:* Follows directly from the regions containing a global optimizer being the only absorbing states, and all other states being transient and leading to an absorbing state.

It is evident that the transition probabilities of the Markov chain depend on the partitioning and the manner by which feasible solutions are generated. This in turn will determine how fast the Markov chain converges to the absorbing state and thus determine the performance of the NP algorithm. It is therefore clearly of great practical importance to develop intelligent partitioning for specific problem structures of interest. The following examples illustrate how to calculate the transition probabilities and how they are dependent on the partitioning.

**Example 2.4** Consider again the example introduced in Example 2.1. Recall that $\sigma_0 = \{1,2,3,4,5,6,7,8\}$ and 14 other valid regions are defined by partitioning each region into 2 subregions ($M_{max} = 2$). This partitioning can be represented by the Markov chain in Figure 2.7, which also shows the transition probabilities given the function values described below.

Further assume that in each iteration, we uniformly generate one point in each region at random ($N_{max} = 1$), and that the relative ranking of the function values is given as follows:

$$f_1 < f_5 < f_6 < f_7 < f_8 < f_3 < f_2 < f_4.$$

To simplify the notation we define $f_i = f(\sigma_i)$, $i = 1, 2, .., 8$. Now the transition probabilities can be easily calculated. See for example, that if the current region is $\sigma_0$, then look at regions $\sigma_1 = \{1,2,3,4\}$, and $\sigma_2 = \{5,6,7,8\}$, and the only way $\sigma_1$ can be found to be better is if point 1 is picked up at random. Since it is assumed that points are selected uniformly, the probability of this is $\frac{1}{4}$. Therefore we get $P_{\sigma_0\ \sigma_1} = \frac{1}{4}$ and it follows that since $P_{\sigma_0\ \sigma_2} = 1 - P_{\sigma_0\ \sigma_1}$, one gets $P_{\sigma_0\ \sigma_2} = \frac{3}{4}$.

Now assume that the current region is $\sigma_1$ and look at the regions $\sigma_3 = \{1,2\}$, $\sigma_4 = \{3,4\}$ and $X \setminus \sigma_1 = \{5,6,7,8\}$. It is clear that no matter which point is selected from $\sigma_4$, one will always generate a better point from $X \setminus \sigma_1$. We can therefore conclude that $P_{\sigma_1\ \sigma_4} = 0$.

Now $\sigma_3$ can only be considered best if solution 1 is randomly selected from that region. Thus $P_{\sigma_1\ \sigma_3} = \frac{1}{2}$. Since there are only three regions that have positive transition probability, the third is now also determined as $P_{\sigma_1\ \sigma_0} = \frac{1}{2}$.

In a similar manner we can for example get $P_{\sigma_3\ \sigma_5} = 1$ and $P_{\sigma_5\ \sigma_5} = 1$. This shows $\sigma_5$ is an absorbing state, which is in agreement with it being the global optimum. The remaining transition probabilities are calculated the

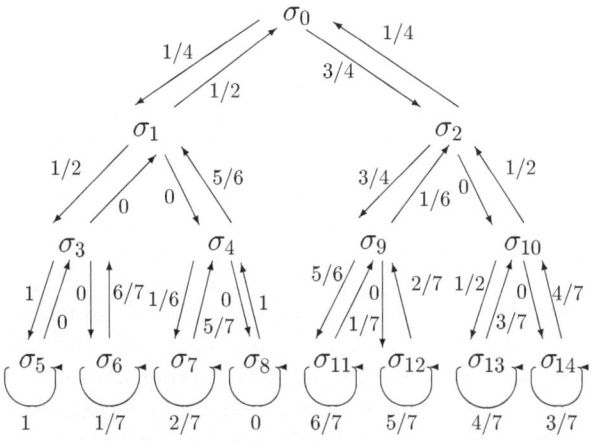

**Fig. 2.7.** Transition probabilities for the Markov chain.

same way, and they are shown in Figure 2.7. It is immediate from the figure that there are no other absorbing states.

In the next example the transition probabilities are calculated for the same problem, but with a different partitioning .

**Example 2.5** Now assume the same problem as in the preceding example, except that in each iteration we partition into 4 subregions ($M_{max} = 4$). Then get the following regions: $\sigma_0 = \{1, 2, 3, 4, 5, 6, 7, 8\}$, $\sigma_1 = \{1, 2\}$, $\sigma_2 = \{3, 4\}$, $\sigma_3 = \{5, 6\}$, $\sigma_4 = \{7, 8\}$, $\sigma_5 = \{1\}$, $\sigma_6 = \{2\}$ and so forth. It is immediately clear that $P_{\sigma_0,\sigma_1} = \frac{1}{2}$ and $P_{\sigma_1,\sigma_5} = 1$.

It can be concluded that there is a 50% probability of getting to the global optimum in only two iterations. This is clearly much better than with the previous partition, and illustrates that how one partitions the feasible region will in general affect how fast the algorithm converges. It should be noted however, that in this second example, 4-5 function evaluations are done in each step that is not at maximum depth, but in the first example, only 2-3 function evaluations were done in each step.

The last two examples clearly illustrate how the partitioning influences the transition probabilities and thus the speed of convergence. It is also clear that the transition probabilities depend on the number of points sampled from each region in each iteration.

**Example 2.6**. Alternatively, it is possible to still partition into two subregions for each region, but to do it in a different manner. Say for example that we let $\eta_1 = \{1, 5, 6, 7\}$, $\eta_2 = \{2, 3, 4, 8\}$, $\eta_3 = \{1, 5\}$, $\eta_4 = \{6, 7\}$, $\eta_5 = \{1\}$ and so forth. Then $P_{\eta_0,\eta_1} = 1$, $P_{\eta_1,\eta_3} = 1$ and $P_{\eta_3,\eta_5} = 1$. Hence the NP method converges to the global optimum in three iterations with probability one. Such *optimal partitioning* always exists but would normally require too much computational effort. On the other hand, it shows how the NP method is heavily dependent on the partitioning and provides motivation for finding intelligent partitioning strategies.

It should be noted that the analysis in this section is identical for integer programming (IP) problems and other problems that may have countable infinite feasible regions, except that the Markov chain now has an infinite state space.

### 2.6.2 Time until Convergence

It has been established that the Markov chain of most promising regions will eventually be absorbed at a global optimum. Although such convergence results are important, it is of high interest to establish bounds on how many iterations, and consequently how many function evaluations, are required before the global optimum is found. In this section such bonds are established for the expected number of iterations until absorption. It is assumed that $\Sigma$

is finite, which is the case for all COPs and for MIPs if the partitioning is defined appropriately. To simplify the analysis, it is assumed that the global optimum is unique.

Let $Y$ denote the number of iterations until the Markov chain is absorbed and let $Y_\eta$ denote the number of iterations spent in state $\eta \in \Sigma$. Since the global optimum is unique, the Markov chain first spends a certain number of iterations in the transient states and when it first hits the unique absorbing state, it never visits any other states. Hence it is sufficient to find the expected number of visits to each of the transient states. Define $T_\eta$ to be the hitting time of state $\eta \in \Sigma$, i.e., the first time that the Markov chain visits the state. Also let $E$ denote an arbitrary event and let $\eta \in \Sigma$ be a valid region. We let $P_\eta[E]$ denote the probability of event $E$ given that the chain starts in state $\eta \in \Sigma$.

It will be convenient to consider the state space $\Sigma$ as consisting of three disjoint subsets as follows: Let $\sigma^*$ be the region corresponding to the unique global optimum. Define $\Sigma_1 = \{\eta \in \Sigma \backslash \{\sigma^*\} \mid \sigma^* \subseteq \eta\}$ and $\Sigma_2 = \{\eta \in \Sigma \mid \sigma^* \not\subseteq \eta\}$. Then $\Sigma = \{\sigma^*\} \cup \Sigma_1 \cup \Sigma_2$ and these three sets are disjoint. Using this notation, we can now state the following result, which relates the expected time to the hitting probabilities.

**Theorem 2.4.** *The expected number of iterations until the NP Markov chain is absorbed is given by*

$$E[Y] = 1 + \sum_{\eta \in \Sigma_1} \frac{1}{P_\eta[T_{\sigma^*} < T_\eta]} + \sum_{\eta \in \Sigma_2} \frac{P_X[T_\eta < \min\{T_X, T_{\sigma^*}\}]}{P_\eta[T_X < T_\eta] \cdot P_X[T_{\sigma^*} < \min\{T_X, T_\eta\}]}.$$
(2.24)

*Proof:* The number of iterations until the Markov chain is absorbed by the global optimum is equal to the number of visits to all the transient states plus one transition into the absorbing state, i.e., $Y = 1 + \sum_{\eta \in \Sigma_1} Y_\eta + \sum_{\eta \in \Sigma_2} Y_\eta$. Hence, since $\Sigma$ is finite and

$$E[Y] = 1 + \sum_{\eta \in \Sigma_1} E[Y_\eta] + \sum_{\eta \in \Sigma_2} E[Y_\eta].$$
(2.25)

We therefore need to show that

$$E[Y_\eta] = \begin{cases} \frac{1}{P_\eta[T_{\sigma^*} < T_\eta]}, & \eta \in \Sigma_1 \\ \frac{P_X[T_\eta < \min\{T_X, T_{\sigma^*}\}]}{P_\eta[T_X < T_\eta] \cdot P_X[T_{\sigma^*} < \min\{T_X, T_\eta\}]}, & \eta \in \Sigma_2. \end{cases}$$
(2.26)

It is known from Markov chain theory that the expected number of visits to transient states is given by the following equation (see Theorem 1 in Hoel, Port and Stone 1972).

$$E[Y_\eta] = \frac{P_X[T_\eta < \infty]}{1 - P_\eta[T_\eta < \infty]}.$$
(2.27)

It is also well known that each transient state is visited finitely many times (Hoel, Port and Stone 1972, p. 19) and since $\Sigma$ is finite, this implies that the hitting time of the absorbing state is finite.

Now first consider the case where $\eta \in \Sigma_1$. For such states, in order to move from the entire feasible region to the global optimum, the Markov chain must first visit $\eta$ at least once. Hence, since the absorbing time is finite, the hitting time of $\eta$ must also be finite, i.e., $P_X[T_\eta < \infty] = 1$. Now if the starting point is $\eta \in \Sigma_1$ then one of two events occurs. Either the Markov chain visits the absorbing state before it visits $\eta$ and thus never visits $\eta$ again, or it visits $\eta$ before it visits the absorbing state. Hence $P_\eta[T_\eta < \infty] = P_\eta[T_{\sigma^*} > T_\eta]$ or equivalently, $1 - P_\eta[T_\eta < \infty] = P_\eta[T_{\sigma^*} < T_\eta]$. This proves the first half of equation (2.26).

Assume that $\eta \in \Sigma_2$. By definition of $\Sigma_2$, $\sigma^* \not\subseteq \eta$ so $\eta \neq X$. Therefore, if the chain starts in state $\eta \in \Sigma_2$ then $T_\eta \neq T_X$ and one of two events occurs: $T_\eta < T_X$ or $T_X < T_\eta$. Hence the probability of finite hitting time can be decomposed as follows:

$$P_\eta[T_\eta < \infty] = P_\eta[T_\eta < T_X] + P_\eta[T_X < T_\eta] \cdot P_X[T_\eta < \infty]$$
$$= 1 - P_\eta[T_X < T_\eta] + P_\eta[T_X < T_\eta] \cdot P_X[T_\eta < \infty],$$

so

$$1 - P_\eta[T_\eta < \infty] = P_\eta[T_X < T_\eta]\left(1 - P_X[T_\eta < \infty]\right). \qquad (2.28)$$

If the chain starts in state $X$ then one of three events occurs, $T_\eta < \min\{T_X, T_{\sigma^*}\}$, $T_X < \min\{T_\eta, T_{\sigma^*}\}$ or $T_{\sigma^*} < \min\{T_X, T_\eta\}$. Since $P_{\sigma^*}(T_\eta < \infty) = 0$, we have

$$P_X[T_\eta < \infty] = P_X[T_\eta < \min\{T_X, T_{\sigma^*}\}] + P_X[T_X < \min\{T_\eta, T_{\sigma^*}\}] \cdot P_X[T_\eta < \infty].$$

This can be rewritten as

$$P_X[T_\eta < \infty] = \frac{P_X[T_\eta < \min\{T_X, T_{\sigma^*}\}]}{1 - P_X[T_X < \min\{T_\eta, T_{\sigma^*}\}]}$$
$$= \frac{P_X[T_\eta < \min\{T_X, T_{\sigma^*}\}]}{P_X[T_\eta < \min\{T_X, T_{\sigma^*}\}] + P_X[T_{\sigma^*} < \min\{T_X, T_\eta\}]} \qquad (2.29)$$

Now we can substitute equation (2.29) into equation (2.28) to obtain

$$1 - P_\eta[T_\eta < \infty] = \frac{P_\eta[T_X < T_\eta] \cdot P_X[T_{\sigma^*} < \min\{T_X, T_\eta\}]}{P_X[T_\eta < \min\{T_X, T_{\sigma^*}\}] + P_X[T_{\sigma^*} < \min\{T_X, T_\eta\}]}.$$

Finally we obtain

$$E[Y_\eta] = \frac{P_X[T_\eta < \infty]}{1 - P_\eta[T_\eta < \infty]} = \frac{P_X[T_\eta < \min\{T_X, T_{\sigma^*}\}]}{P_\eta[T_X < T_\eta] \cdot P_X[T_{\sigma^*} < \min\{T_X, T_\eta\}]}.$$

This completes the proof.

The expected number of iterations (2.24) depends on both the partitioning and the manner by which feasible sample solutions are generated, as well as the structure of the problem itself. Calculating this expectation exactly is therefore complicated. However, with some additional assumptions it is possible to find useful bounds on the expected number of iterations.

Assume that $P^*$ is a lower bound on the probability of selecting the correct region. We refer to the probability $P^*$ as the minimum success probability. The next theorem provides an upper bound for the expected time until the NP algorithm converges in terms of $P^*$ and the size of the feasible region as measured by $d^*$.

**Theorem 2.5.** *Assume that $P^* > 0.5$. The expected number of iterations until the NP Markov chain is absorbed is bounded by*

$$E[Y] \le \frac{d^*}{2P^* - 1}. \tag{2.30}$$

*Proof:* By definition of $P^*$ the Markov chain $\{\sigma(k)\}$ has a minimum success probability of $P^*$ given its current state $\sigma(k) \in \Sigma$, that is, with probability of at least $P^*$, $\sigma(k+1)$ will be closer to $\sigma^*$ than $\sigma(k)$ in terms of the number of transitions required to move between the regions. Now imagine a Markov chain that is identical to $\{\sigma(k)\}$ except that this success probability is even and equal to $P^*$ for every state $\sigma \in \Sigma$. Now note that since the success probability is constant, the exact state is not of any consequence, but rather the number of transitions it takes to move from the current state $\sigma(k)$ to the optimum. The maximum such distance is $2d^*$, and we can therefore, without losing any information, reduce the state space to $\mathcal{S} = \{0, 1, 2, ...., 2d^*\}$. With this representation the entire feasible region $X$ corresponds to state $d^*$, and we can let the global optimum correspond to state zero. Given a state $x \in \mathcal{S}$ the probability of moving to $x - 1$ is fixed and equal to $P^*$, and the probability of moving to $x + 1$ is equal to $1 - P^*$, independent of the state. Therefore, the new Markov chain is a simple random walk. Furthermore, it is clear that $E[Y] \le E[Y']$, where $Y'$ is the first time the random walk visits $\sigma^*$ if it starts in state $d^*$, which corresponds to $X$ (the starting state of the NP Algorithm).

Hence, if we calculate the expected hitting time for the random walk this automatically gives us an upper bound for the original Markov chain. Furthermore, since we are only interested in the first time the global optimum is encountered, we can assume 0 is an absorbing barrier and look at the time of absorption. Note also that $2d^*$ is a reflective barrier. Then it is known that the expected time $T$ of absorption when starting in state $u$ is (Weesakul, 1961)

$$E_u[T] = \frac{u}{2P^* - 1} + \frac{(1 - P^*)^{2d^*+1}}{(P^*)^{2d^*}(2P^* - 1)^2} \left(1 - \left(\frac{P^*}{1 - P^*}\right)^u\right). \tag{2.31}$$

Thus, for $u = d^*$, that is, when the algorithm starts in state $\sigma(0) = X$, we have

$$E_{d^*}[T] = \frac{d^*}{2P^* - 1} + \frac{(1 - P^*)^{2d^*+1}}{(P^*)^{2d^*}(2P^* - 1)^2}\left(1 - \left(\frac{P^*}{1 - P^*}\right)^{d^*}\right) \equiv C_1. \quad (2.32)$$

Now since $P^* > 0.5$, $\frac{P^*}{1-P^*} > 1$, so

$$\left(1 - \left(\frac{P^*}{1 - P^*}\right)^{d^*}\right) < 0,$$

and hence

$$\frac{(1 - P^*)^{2d^*+1}}{(P^*)^{2d^*}(2P^* - 1)^2}\left(1 - \left(\frac{P^*}{1 - P^*}\right)^{d^*}\right) < 0.$$

Therefore,

$$E[Y] \leq E_{d^*}[T] \leq \frac{d^*}{2P^* - 1},$$

which proves the theorem.

Note that (2.30) grows only linearly in $d^*$. Furthermore, as for any other tree, the depth $d^*$ of the partitioning tree is a logarithmic function of the input parameter(s). In other words, assume for simplicity that there is a single input parameter $n$. Then (2.30) provides a bound on the expected number of iterations that is $O(\log n)$. This shows that the expected number of iterations required by the NP algorithm grows very slowly in the size of the problem, which partially explains why the NP algorithm is very effective for solving large-scale optimization problems.

On the other hand, the bounds (2.30) grow exponentially as $P^* \to \frac{1}{2}$. This is further illustrated in Table 2.1, which shows the bounds for several problem sizes ($d^*$) and values for the minimum success probability ($P^*$). Clearly the

**Table 2.1.** Bounds on the expected time until convergence.

| Success Prob. | Maximum Depth ($d^*$) | | | | |
|---|---|---|---|---|---|
| | 2 | 5 | 10 | 20 | 30 |
| 55% | 20 | 50 | 100 | 200 | 300 |
| 60% | 10 | 25 | 50 | 100 | 150 |
| 65% | 7 | 17 | 33 | 67 | 100 |
| 70% | 5 | 13 | 25 | 50 | 75 |
| 75% | 4 | 10 | 20 | 40 | 60 |
| 80% | 3 | 8 | 17 | 33 | 50 |
| 85% | 3 | 7 | 14 | 29 | 43 |
| 90% | 3 | 6 | 13 | 25 | 38 |
| 95% | 2 | 6 | 11 | 22 | 33 |

number of expected iterations increases rapidly as the success probability decreases.

These results underscore the previously made statement that the efficiency of the NP algorithm depends on making the correct move frequently. This success probability depends in turn on both the partitioning and the method for generating feasible solutions. For any practical application it is therefore important to increase the success probability by developing intelligent partitioning methods, incorporating special structure into weighted sampling, and applying randomized heuristics to generate high-quality feasible solutions.

## 2.7 Continuous Optimization Problems

Although the main focus of this book is on discrete optimization problems, it should be noted that the NP method could be applied to continuous optimization problems and Mixed Integer Programs (MIP) where the feasible region is uncountable but bounded, similar analysis can also be done but some additional conditions are required.

We now turn our attention to the case where $X$ is not countable, but a bounded subset of $\mathbf{R}^n$. If $X$ is not convex and $f$ is neither differentiable nor convex, then conventional mathematical programming techniques usually cannot guarantee convergence to a global optimum. Many global optimization methods have been suggested for this difficult problem and we refer the reader to Horst and Pardalos (1995), and Törn and Žilinskas (1989) for a review. However, the NP method also provides an attractive alternative for the problem and hence it is worthwhile to consider how it may be applied.

It is not difficult to verify that the NP method generates a Markov chain when applied to the optimization problem with uncountable but bounded feasible region. It is also easy to see that the only states containing a global optimum can be absorbing states. However, some additional conditions are needed to prove that all such states will be absorbing.

Since $X$ is not countable, we can partition the solution space such that the regions at maximum depth are arbitrarily small. In this case the maximum depth is determined by the desired accuracy. On the other hand, by selecting an appropriate promising index, global convergence can be assured for this problem.

Let $H : X \times \Sigma \to X$ denote a local search heuristic that is constrained to stay within a certain valid region. For example, for any $\sigma \in \Sigma$, if $x \in \sigma$, then the outcome of $H(x, \sigma)$ is a point, $\tilde{x}$, in $X$ that the local search converges to with $x$ as the starting point. Here, $\tilde{x}$ is constrained in $\sigma$. If $x \notin \sigma$, we can define $H(x, \sigma) = x$. With respect to $H$, each local minimum has a region of attraction that is defined as follows: For a point $\tilde{x} \in \mathbf{R}^n$ the region of attraction is the set $A(\tilde{x}) \subseteq \mathbf{R}^n$ such that for any $x \in A(\tilde{x})$, $H(x, X) = \tilde{x}$. We can now define a promising index as

$$I(\sigma) = \min_{x \in \sigma} f\left(H(x, \sigma)\right). \tag{2.33}$$

We refer to this promising index as the *descent promising index*. Given a sample of $N$ points $D_\sigma = \{x_k\}_{k=1}^N$ from a region $\sigma$. The descent promising index can be estimated using

$$\hat{I}(\sigma) = \min\left\{f(H\left(x_1, \sigma\right)), ..., f(H\left(x_N, \sigma\right))\right\}. \tag{2.34}$$

Notice that for the descent promising index it is often reasonable to use a small sample size, since many of the sample points are likely to converge to the same point.

We finally note that when applying the NP method to continuous optimization, we implicitly assume that $X$ can be partitioned and sampled.

Similar to above, we can verify that the NP method generates a Markov chain when applied to the optimization problem with uncountable but bounded feasible region. We then prove that the states corresponding to a global optimum are the only absorbing states. This establishes global convergence. For simplicity we assume here that the global optimum is unique.

The essential condition for global convergence is that the maximum depth regions are small enough. To make this rigorous, define a function, $\phi : X \to \mathbf{R}$, which measures the distance between a point $\tilde{x} \in \mathbf{R}^n$ and the compliment of its region of attraction,

$$\phi(\tilde{x}) = \inf_{x \in X \setminus A(\tilde{x})} ||x - \tilde{x}||, \tag{2.35}$$

where $|| \cdot ||$ is the Euclidean norm.

**Theorem 2.6.** *Assume the NP method is applied to a problem defined on a bounded subset $X$ of $\mathbf{R}^n$ using the descent promising index defined by equation (2.33). Choose the maximum depth $d^*$ such that it satisfies*

$$\phi(x^*) > \max_{x_1, x_2 \in \eta} ||x_1 - x_2|| \tag{2.36}$$

*for all $\eta \in \Sigma_0$. Let $\sigma_{opt}$ be the maximum depth region that contains the optimum solution $x^*$, then $\sigma_{opt}$ is an absorbing state and the NP method converges to the global optimum in finite time with probability one.*

*Proof:* Let $x \in \sigma_{opt}$. Due to the condition (2.36) and the definition of the $\phi$, we have that $H(x, \sigma_{opt}) = \theta_{opt}$. Hence, the estimate of the promising index is $\hat{I}(\sigma_{opt}) = f(x^*)$ regardless of which sample points are selected. It follows that $\hat{I}(\sigma_{opt}) < \hat{I}(\eta)$ for all other valid regions $\eta \in \Sigma$ and hence $\sigma_{opt}$ is an absorbing state. It is straightforward to see that no other states can be absorbing.

# 3

# Noisy Objective Functions

In this chapter we develop the theory of the NP method for problems where the objective function is noisy, for example when it can only be evaluated as a realization of some random variable. This complicates the objective function and occurs in numerous contexts where a simple analytical model cannot be specified for the system being optimized. For example, the system may be modeled by a discrete event simulation model and the objective function can then only be estimated as the output of the simulation model. The data mining and radiation treatment problems introduced in Chapter 1 provide two more examples. In the data mining model it may be necessary to evaluate the performance with only a fraction of the data since the amount of data is often too large to work with all at once. This introduces noise into the performance. For the radiation treatment problem the objective is evaluated by outside experts. Different experts may evaluate the same radiation treatment plan differently and the same expert may even have different evaluations at different times. Again, the objective function therefore becomes noisy.

Problems with noisy objective functions add some special considerations from both practical and theoretical points of view. However, some implementation aspects remain unchanged and most importantly the partitioning is not affected by noise in the objective function. Even if when we use an intelligent partitioning that incorporates the objective function in some manner, estimates can be used directly instead of exact values and the implementation is essentially unchanged. All changes in the implementation and analysis of the NP algorithm thus deal with how the algorithm moves on the fixed state space.

First, from a practical implementation point of view there is now additional noise introduced into the decision of determining which region should become the most promising region in the next iteration. Everything else being equal, this can be expected to reduce the probability that the correct region is selected. The question of determining how much computational effort is needed to make the correct selection is therefore more pertinent than ever. Furthermore, the noise is often controllable to some extent by using more computation effort. For example, if the noise is due to the evaluation being based on a simulation model then increasing the simulation run length or number of runs will decrease the noise. A practical question thus becomes how much effort should be devoted to evaluating the performance.

From a more theoretical point of view it is now clear that the NP method no longer generates an absorbing Markov chain, which was the basis for finite time convergence of the deterministic NP method. The reason for this is that even though a singleton region corresponding to a global optimum is visited it is now possible that due to the noise the NP method will leave this singleton. It is thus necessary to revisit the convergence of the NP method and it is unlikely that finite time convergence could ever be assured for a problem with noisy performance.

In this chapter we first establish some basic properties and asymptotic convergence of the NP method for noisy performance. We then discuss some important implementation issues regarding how to generate sample solutions such that the correct selection of the next most promising region is assured with a sufficient probability. Finally, we discuss the finite time behavior of the method.

## 3.1 Convergence Analysis

Since the NP method uses the set performance function $I: \Sigma \to \mathbf{R}$ to guide its movements we shift our focus to finding an element $\sigma^* \in \mathcal{S}$, where

$$\mathcal{S} = \arg \min_{\sigma \in \Sigma_0} I(\sigma). \tag{3.1}$$

Given certain assumptions, we show that the NP method identifies an element in $\mathcal{S}$, that is, an optimal solution for the problem (3.1) above. This, however, is equivalent to solving the original problem since the original performance function $f: X \to \mathbf{R}$ and the set performance function $I: \Sigma \to \mathbf{R}$ agree on singletons.

### 3.1.1 Basic Properties

We start by showing that the NP method generates an ergodic Markov chain. Furthermore, we show that the estimate of the best singleton region converges with probability one to a maximizer of the stationary distribution of the Markov chain. The stationary distribution can thus be used as a basis for inference. For any $\sigma, \eta \in \Sigma$, we use $P(\sigma, \eta)$ to denote the transition probability of the Markov chain moving from state $\sigma$ to state $\eta$, and $P^n(\sigma, \eta)$ to denote the $n$-step transition probability. We also let $\mathcal{P}[A]$ denote the probability of any event $A$.

Since in the $k$th iteration, the next most promising region $\sigma(k + 1)$ only depends on the current most promising region $\sigma(k)$, and the sampling information obtained in the $k$th iteration, the sequence of most promising regions

$\{\sigma(k)\}_{k=1}^{\infty}$ is a Markov chain with state space $\Sigma$. We refer to this Markov chain as the *NP Markov chain*.

In the analysis below we let $\Sigma_0 \subset \Sigma$ denote all of the maximum depth (singleton) regions. We also let $\mathcal{H}(\sigma)$ denote the set of subregions of a valid region $\sigma \in \Sigma \setminus \Sigma_0$, and $b(\sigma)$ denote the parent region of each valid region $\sigma \in \Sigma \setminus \{X\}$.

In each iteration the NP method selects the next most promising region based on which one has the best estimated promising index. Therefore, the transition probabilities may be written in terms of the estimated promising index. We need to consider three cases:

1. If $\sigma = X$, then the NP algorithm moves to one of the subregions in the next iteration and the set of subregions that have the best estimated performance is given by

$$B_1 = \left\{\xi \in \mathcal{H}(\sigma) | \hat{I}(\xi) \le \hat{I}(\eta), \forall \eta \in \mathcal{H}(\sigma)\right\}. \tag{3.2}$$

The transition probabilities are given by

$$P(\sigma, \xi) = \sum_{i=1}^{|\mathcal{H}(\sigma)|} \frac{\mathcal{P}\left[\xi \in B_1 : |B_1| = i\right] \cdot \mathcal{P}[|B_1| = i]}{i} \tag{3.3}$$

for $\xi \in \mathcal{H}(\sigma)$ and $P(\sigma, \xi) = 0$ for $\xi \notin \mathcal{H}(\sigma)$. Note that this equation assumes that ties are broken in a uniform manner.

2. If $\sigma \in \Sigma \setminus (\{X\} \cup \Sigma_0)$, then the NP algorithm either moves to a subregion or backtracks in the next iteration and the set of regions that have the best estimated performance is given by

$$B_2 = \left\{\xi \in \mathcal{H}(\sigma) \cup \{X \setminus \sigma\} | \hat{I}(\xi) \le \hat{I}(\eta), \forall \eta \in \mathcal{H}(\sigma) \cup \{X \setminus \sigma\}\right\}. \tag{3.4}$$

The transition probabilities are thus given by

$$P(\sigma, \xi) = \sum_{i=1}^{|\mathcal{H}(\sigma)|+1} \frac{\mathcal{P}\left[\xi \in B_2 : |B_2| = i\right] \cdot \mathcal{P}[|B_2| = i]}{i} \tag{3.5}$$

for $\xi \in \mathcal{H}(\sigma) \cup \{X \setminus \sigma\}$ and $P(\sigma, \xi) = 0$ for $\xi \notin \mathcal{H}(\sigma) \cup \{X \setminus \sigma\}$.

3. If $\sigma \in \Sigma_0$, then the NP algorithm either stays in the current most promising region or backtracks in the next iteration, and the transition probabilities are given by

$$P(\sigma, \xi) = \begin{cases} \mathcal{P}\left[\hat{I}(X \setminus \sigma) < \hat{I}(\sigma)\right] + \frac{\mathcal{P}[\hat{I}(X \setminus \sigma) = \hat{I}(\sigma)]}{2}, & \xi = b(\sigma) \\ \mathcal{P}\left[\hat{I}(X \setminus \sigma) > \hat{I}(\sigma)\right] + \frac{\mathcal{P}[\hat{I}(X \setminus \sigma) = \hat{I}(\sigma)]}{2}, & \xi = \sigma \\ 0, & \text{otherwise.} \end{cases} \tag{3.6}$$

Equations (3.2) - (3.6) completely define the transition probabilities for the NP algorithm.

We will show that the NP Markov chain has a unique stationary distribution, but first we exclude certain trivial cases from the analysis. In particular we assume that there is no region $\sigma \in \Sigma \setminus \{X\}$ such that despite noisy estimates, every point in $\sigma$ has, with probability one, better estimated performance than all of the points outside of $\sigma$, that is in $X \setminus \sigma$. Formally we state the assumption as follows:

**Assumption 3.1** *For all* $\sigma \in \Sigma \setminus \{X\}$, $P(\sigma, b(\sigma)) > 0$.

We note that this assumption implies that for any $\eta \in \Sigma$, $P^{d(\eta)}(\eta, X) > 0$. Also, since the random sampling scheme assumes that each point in the sample region is sampled with positive probability, Assumption 1 is equivalent to assuming that for all valid regions $\sigma \in \Sigma$ there exists at least one pair of points $x_1 \in \sigma$ and $x_2 \in X \setminus \sigma$ such that $\mathcal{P}[\hat{f}(x_1) > \hat{f}(x_2)] > 0$. This observation shows that the assumption imposes no loss of generality. Indeed, if it does not hold, then we can let $\sigma_0$ be the smallest region for which it is not satisfied. Then it is straightforward to verify that $\sigma_0$ is closed, all the other states are transient, and the Markov chain eventually enters $\sigma_0$ and never leaves. Thus, $\sigma_0$ contains all the interesting solutions, since, with probability one, all the points in $\sigma_0$ have better performance than all points outside $\sigma_0$. If $\sigma_0$ is a singleton it is an absorbing state which the NP method converges to in finite time, and no further analysis is necessary. If $\sigma_0$ is not a singleton the analysis below can be applied by replacing $X$ with $\sigma_0$ as the feasible region. This assumption thus imposes no loss of generality.

**Proposition 3.1.** *If Assumption 3.1 holds, then the NP Markov chain has a unique stationary distribution* $\{\pi(\sigma)\}_{\sigma \in \Sigma}$.

*Proof:* Let $\Sigma_P$ denote the set of positive recurrent states of the Markov chain. Since $\Sigma$ is finite $\Sigma_P \neq \emptyset$, which implies that the NP Markov chain has at least one stationary distribution. To show that it is also unique, we must show that $\Sigma_P$ is irreducible. We first show that $X \in \Sigma_P$. Let $\sigma \in \Sigma_P$. Then by Assumption 3.1 there exists some $k$ such that $P^k(\sigma, X) > 0$ and since $\sigma$ is positive recurrent $X$ is also positive recurrent. Now to prove irreducibility let $\sigma, \eta \in \Sigma_P$. Again by Assumption 3.1 there exists some $k_1$ and $k_2$ such that $P^{k_1}(\sigma, X) > 0$ and $P^{k_2}(\eta, X) > 0$. Furthermore, since $\eta \in \Sigma_P$ is recurrent and $P^{k_2}(\eta, X) > 0$ then there exists some $k_3$ such that $P^{k_3}(X, \eta) > 0$. Hence $P^{k_1+k_3}(\sigma, \eta) > 0$. This holds for any $\sigma, \eta \in \Sigma_P$, so $\Sigma_P$ is irreducible, and thus the Markov chain has a unique stationary distribution $\{\pi(\sigma)\}_{\sigma \in \Sigma}$.

We note that since a unique stationary distribution exists, the average number of visits to each state converges to this distribution with probability one (w.p.1), that is,

$$\lim_{k \to \infty} \frac{\mathcal{N}_k(\sigma)}{k} = \pi(\sigma), \ \forall \sigma \in \Sigma \text{ w.p.1}. \tag{3.7}$$

Furthermore, as $P(\sigma, \sigma) > 0$ for some $\sigma \in \Sigma_0$, the chain is aperiodic and hence the $k$-step transition probabilities converge pointwise to the stationary distribution, $\lim_{k \to \infty} P^k(X, \sigma) = \pi(\sigma)$ w.p.1, $\forall \sigma \in \Sigma$. The next step in establishing the convergence of the NP method is to use (3.7) to show that the estimate of the best solution converges to a maximum stationary probability of the NP Markov chain and hence that the stationary distribution can be used for inference.

**Theorem 3.2.** *Assume that Assumption 3.1 holds. The estimate of the best region $\hat{\sigma}^*(k)$ converges to a maximum of the stationary distribution of the NP Markov chain, that is,*

$$\lim_{k \to \infty} \hat{\sigma}^*(k) \in \arg\max_{\sigma \in \Sigma_0} \pi(\sigma), \quad w.p.1. \tag{3.8}$$

*Proof:* Let $k_0 = \min_{k \geq 1} \{k | \sigma(k) \in \Sigma_0\}$ be the first time the NP method reaches maximum depth. If $\sigma^* \in \mathcal{S}$ is a global optimum, then there is a positive probability that $\sigma^*$ is sampled for $d(\sigma^*)$ consecutive iterations, which together with $X \in \Sigma_P$ implies that $P^{d(\sigma^*)}(X, \sigma^*) > 0$, i.e., $\sigma^* \in \Sigma_P$. Thus there is at least one positive recurrent state in $\Sigma_0$ and consequently $k_0 < \infty$ almost surely.

Now combine this with the known fact that equation (3.7) holds and let $A$ be such a set that $\mathcal{P}[A] = 1$ and for all $\omega \in A$, $k_0(\omega) < \infty$ and $\lim_{k \to \infty} \frac{\mathcal{N}_k(\sigma, \omega)}{k} = \pi(\sigma)$. Let $\omega \in A$ be a fixed sample point. Since $\Sigma_0$ is finite there exists a finite constant $K_\omega \geq k_0(\omega)$ such that for all $k \geq K_\omega$,

$$\arg\max_{\sigma \in \Sigma_0} \mathcal{N}_k(\sigma) \subseteq \arg\max_{\sigma \in \Sigma_0} \pi(\sigma).$$

Now by definition $\hat{\sigma}^*(k, \omega) \in \arg\max_{\sigma \in \Sigma_0} \mathcal{N}_k(\sigma, \omega)$ for all $k \geq k_0$, so $\hat{\sigma}^*(k, \omega) \in \arg\max_{\sigma \in \Sigma_0} \pi(\sigma)$ for all $k \geq K_\omega$. This completes the proof.

To prove asymptotic global convergence we need to show that the singleton maximizer of the stationary distribution can be used as an estimate of the global optimum. As will be explained shortly this may not hold without some conditions that can be satisfied by an effective means of partitioning and generating feasible solutions.

### 3.1.2 Global Convergence

We now show that given certain regularity conditions, the observation that the NP method converges to a maximizer of the stationary distribution implies that it converges to a global optimum. In particular, by showing that the maximum depth region with the largest stationary probability is also a global optimum, it follows that the NP method converges to the global optimum.

Without assumptions about the problem structure this does not necessarily hold. For example, assume that $\sigma^*$ is a unique global optimum, but that there exists a set $\eta \in \Sigma$ such that $\sigma^* \subseteq \eta$ and $J(x) > J(\phi)$ for all $x \in \eta \setminus \sigma^*$,

$\phi \in \mathbf{X} \setminus \eta$. Then we expect the transition probability $P^{d(\eta)}(\mathbf{X}, \eta)$ into $\eta$ to be small, which implies that $P^{d(\sigma^*)}(\mathbf{X}, \sigma^*)$ into $\sigma^*$ is also likely to be small. This indicates that $\pi(\sigma^*)$ may be quite small. Therefore, if the performance measure changes drastically from the global optimum to points that are close in the partitioning metric, the NP method is not guaranteed to find the global optimum. Consequently, we must exclude such problem structures. We start with a definition.

**Definition 3.3.** *Let $\sigma$ and $\eta$ be any valid regions. Then there exists some sequence of regions $\sigma = \xi_0, \xi_1, ..., \xi_n = \eta$ along which the Markov chain can move to get from state $\sigma$ to state $\eta$. We call the shortest such sequence the shortest path from $\sigma$ to $\eta$. We also define $\kappa(\sigma, \eta)$ to be the length of the shortest path.*

Note that the shortest path is unique and can be found by moving directly from $\sigma$ to the smallest valid region containing both $\sigma$ and $\eta$, and then by proceeding directly from this region to $\eta$. The following condition on transition probabilities along shortest paths connecting the global optimum to other maximum depth regions turns out to be a sufficient condition for global convergence. The interpretation of this assumption is discussed further at the end of the section.

**Assumption 3.2** *The set*

$$\mathcal{S}_0 = \left\{ \xi \in \Sigma_0 : P^{\kappa(\eta, \xi)}(\eta, \xi) \geq P^{\kappa(\xi, \eta)}(\xi, \eta), \forall \eta \in \Sigma_0 \right\} \tag{3.9}$$

*satisfies $\mathcal{S}_0 \subseteq \mathcal{S}$, that is, it is a subset of the set of global optimizers.*

The regularity condition in equation (3.9), needed for global convergence of both the NP I and NP II algorithms, guarantees that the transition probability from any part of the feasible region to the global optimum is at least as large as going from the global optimum back to that region. This ensures that the points close to the global optimum are sufficiently good. We now use this assumption to prove global convergence of the NP I algorithm.

**Theorem 3.4.** *Assume that the NP I algorithm is applied and Assumptions 3.1-3.2 hold. Then*

$$\arg \max_{\sigma \in \Sigma_0} \pi(\sigma) \subseteq \mathcal{S}, \tag{3.10}$$

*and consequently the NP I algorithm converges with probability one to a global optimum.*

*Proof:* Let $\eta_1, \eta_2$ be any regions of maximum depth. Let $\xi_0, \xi_1, ..., \xi_{n-1}, \xi_n$ be the shortest path from $\eta_1$ to $\eta_2$, where $n = \kappa(\eta_1, \eta_2)$. It is also clear from the Kolmogorov criterion (Wolff 1989) that the NP I Markov chain is reversible, and we get

$$P(\xi_i, \xi_{i+1}) \pi(\xi_i) = P(\xi_{i+1}, \xi_i) \pi(\xi_{i+1}), \quad i = 0, 1, 2, ..., n - 1.$$

We can rewrite this as

$$\pi(\xi_i) = \frac{P(\xi_{i+1}, \xi_i)}{P(\xi_i, \xi_{i+1})}\pi(\xi_{i+1}), \ \ i = 0, 1, 2, ..., n - 1.$$

By using this equation iteratively and using $\xi_0 = \eta_1$ and $\xi_n = \eta_2$ we can rewrite this as

$$\pi(\eta_1) = \frac{P(\xi_n, \xi_{n-1})...P(\xi_2, \xi_1)P(\xi_1, \xi_0)}{P(\xi_0, \xi_1)P(\xi_1, \xi_2)...P(\xi_{n-1}, \xi_n)}\pi(\eta_2). \tag{3.11}$$

Furthermore, since $\xi_0, \xi_1, ..., \xi_{n-1}, \xi_n$ is the shortest path from $\xi_0$ to $\xi_n$ then we also have that

$$P^n(\eta_1, \eta_2) = P^n(\xi_0, \xi_n) = P(\xi_0, \xi_1)P(\xi_1, \xi_2)...P(\xi_{n-1}, \xi_n),$$

and

$$P^n(\eta_2, \eta_1) = P^n(\xi_n, \xi_0) = P(\xi_n, \xi_{n-1})...P(\xi_2, \xi_1)P(\xi_1, \xi_0).$$

We can therefore rewrite equation (3.11) as

$$\pi(\eta_1) = \frac{P^n(\eta_2, \eta_1)}{P^n(\eta_1, \eta_2)}\pi(\eta_2) = \frac{P^{\kappa(\eta_2, \eta_1)}(\eta_2, \eta_1)}{P^{\kappa(\eta_1, \eta_2)}(\eta_1, \eta_2)}\pi(\eta_2). \tag{3.12}$$

Now if $\eta_2 \in \arg\max_{\sigma \in \Sigma_0} \pi(\sigma)$ then by definition $\pi(\eta_1) \leq \pi(\eta_2)$ so by equation (3.12) we have $\frac{P^{\kappa(\eta_2, \eta_1)}(\eta_2, \eta_1)}{P^{\kappa(\eta_1, \eta_2)}(\eta_1, \eta_2)} \leq 1$, that is, $P^{\kappa(\eta_1, \eta_2)}(\eta_1, \eta_2) \geq P^{\kappa(\eta_2, \eta_1)}(\eta_2, \eta_1)$. Since $\eta_1$ can be chosen arbitrarily, Assumption 4.6 now implies that $\eta_2 \in \mathcal{S}_0 \subseteq \mathcal{S}$, which proves the theorem.

Similarly, the following theorem establishes global convergence for the NP II algorithm.

**Theorem 3.5.** *Assume that the NP II algorithm is applied and Assumptions 3.1-3.2 hold. Then*

$$\arg\max_{\sigma \in \Sigma_0} \pi(\sigma) \subseteq \mathcal{S}, \tag{3.13}$$

*and consequently the NP II algorithm converges with probability one to a global optimum.*

*Proof:* We start by looking at a state $\eta \in \Sigma \setminus (\Sigma_0 \cup X)$. Since the chain leaves this state with probability one, and the state can only be entered by transitions from state $s(\eta)$, the balance equations are given by

$$\pi(\eta) = P(s(\eta), \eta)\pi(s(\eta)).$$

These equations can be solved iteratively to obtain

$$\pi(\eta) = P^{d(\eta)}(X, \eta)\pi(X). \tag{3.14}$$

Now assume $\sigma \in \Sigma_0$. The balance equations are given by

$$\pi(\sigma)P(\sigma, X) = \pi(s(\sigma))P(s(\sigma), \sigma).$$

Rewrite and use equation (3.14) above to get

$$
\begin{aligned}
\pi(\sigma) &= \frac{P(s(\sigma), \sigma))}{P(\sigma, X)}\pi(s(\sigma)) \\
&= \frac{P(s(\sigma), \sigma))}{P(\sigma, X)}P^{d(s(\sigma))}(X, s(\sigma))\pi(X) \\
&= \frac{P^{d(\sigma)}(X, \sigma)}{P(\sigma, X)}\pi(X).
\end{aligned}
\tag{3.15}
$$

Now take another $\eta \in \Sigma_0$, then

$$
\begin{aligned}
\pi(\sigma) &= \frac{P^{d(\sigma)}(X, \sigma)}{P(\sigma, X)}\pi(X) \\
&= \frac{P^{d(\sigma)}(X, \sigma)}{P(\sigma, X)}\frac{P(\eta, X)}{P^{d(\eta)}(X, \eta)}\pi(\eta) \\
&= \frac{P^{d(\sigma)+1}(\eta, \sigma)}{P^{d(\eta)+1}(\sigma, \eta)}\pi(\eta) \\
&= \frac{P^{\kappa(\eta, \sigma)}(\eta, \sigma)}{P^{\kappa(\sigma, \eta)}(\sigma, \eta)}\pi(\eta).
\end{aligned}
$$

The remainder of the proof follows the last paragraph of Theorem 3.4 above and is omitted.

We note that a sufficient condition for (3.9) to be satisfied is that in each iteration, the probability of making the 'correct' decision is at least one half, and that the condition (3.9) depends on the partitioning, sampling, and method of estimation. This condition needed for global convergence is in general difficult to verify, but to gain some insights the remainder of this section is devoted to obtaining conditions on the partitioning , sampling, and estimation elements that are sufficient for convergence. These conditions are considerably stronger than (3.9) but are somewhat more intuitive. For simplicity we assume that there is a unique global optimum $\sigma^* \in \mathcal{S}$. We start with a definition.

**Definition 3.6.** *Let* $\Sigma_g = \{\sigma \in \Sigma : \sigma^* \subseteq \sigma\}$ *denote all the regions containing the unique global optimum* $\sigma^*$ *and* $\Sigma_b = \Sigma \setminus \Sigma_g$ *denote the remaining regions. Define a function* $Y : \Sigma_g \times \Sigma_b \to \mathbf{R}$ *by*

$$Y(\sigma_g, \sigma_b) = J\left(x_{\sigma_b}^{[1]}\right) - J\left(x_{\sigma_g}^{[1]}\right), \tag{3.16}$$

*where* $x_\sigma^{[1]}$ *denotes the random sample point from* $\sigma \in \Sigma$ *that has the estimated best performance and is generated in Step 2 of the NP algorithm. Furthermore, let*

$$\hat{Y}(\sigma_g, \sigma_b) = L\left(x_{\sigma_b}^{[1]}\right) - L\left(x_{\sigma_g}^{[1]}\right) \tag{3.17}$$

*denote the corresponding simulation estimate.*

We note that for each $\sigma \in \Sigma$, $x_\sigma^{[1]} \in \sigma$ is a random variable that is defined by the sampling strategy, and that

$$L\left(x_\sigma^{[1]}\right) = \hat{I}(\sigma). \tag{3.18}$$

Also note that for any $\sigma_g \in \Sigma_g$, $\sigma_b \in \Sigma_b$, the random variable $Y(\sigma_g, \sigma_b)$ is defined by the partitioning and sampling strategies, whereas if we condition the random variable $\hat{Y}(\sigma_g, \sigma_b)$ on the value of $Y(\sigma_g, \sigma_b)$, its outcome depends only on the estimated sample performance, that is, on the randomness inherent in the simulation estimate. We now obtain the following theorem for the NP I algorithm:

**Theorem 3.7.** *Assume that $\sigma^*$ is unique and*

$$\mathcal{P}\left[\hat{Y}(\sigma_g, \sigma_b) > 0 | Y(\sigma_g, \sigma_b) = y\right] > \frac{y}{2E[Y(\sigma_g, \sigma_b)]} \tag{3.19}$$

*for all $\min_{x \in \sigma_b} J(x) - \max_{x \in \sigma_g} J(x) \le y \le \max_{x \in \sigma_b} J(x) - \min_{x \in \sigma_g} J(x)$, $\sigma_g \in \Sigma_g$, and $\sigma_b \in \Sigma_b$. Then the NP I algorithm converges to a global optimum.*

*Proof:* First note that the assumption of this theorem can be satisfied only if

$$\frac{y}{2E[Y(\sigma_g, \sigma_b)]} < 1,$$

where we have $y \le \max_{x \in \sigma_b} J(x) - \min_{x \in \sigma_g} J(x)$. Thus, we must have

$$E[Y(\sigma_g, \sigma_b)] > \frac{1}{2}\left(\max_{x \in \sigma_b} J(x) - \min_{x \in \sigma_g} J(x)\right) = \frac{1}{2}\left(\max_{x \in \sigma_b} J(x) - J(x_{opt})\right), \tag{3.20}$$

and thus $E[Y(\sigma_g, \sigma_b)] > 0$. Now let $\sigma_g \in \Sigma_g, \sigma_b \in \Sigma_b$, let $F_Y \equiv F_{Y(\sigma_g, \sigma_b)}$ denote the distribution function of $Y(\sigma_g, \sigma_b)$, and condition on the value of $Y(\sigma_g, \sigma_b)$:

$$\mathcal{P}\left[\hat{Y}(\sigma_g, \sigma_b) > 0\right] = \int_{\mathbf{R}} \mathcal{P}\left[\hat{Y}(\sigma_g, \sigma_b) > 0 | Y(\sigma_g, \sigma_b) = y\right] dF_Y(y)$$

$$> \int_{\mathbf{R}} \frac{y}{2E[Y(\sigma_g, \sigma_b)]} dF_Y(y)$$

$$= \frac{1}{2E[Y(\sigma_g, \sigma_b)]} \int_{\mathbf{R}} y dF_Y(y)$$

$$= \frac{1}{2}.$$

Now since by definition $\hat{Y}(\sigma_g, \sigma_b) = L\left(x_{\sigma_b}^{[1]}\right) - L\left(x_{\sigma_g}^{[1]}\right)$, this implies that

$$\mathcal{P}\left[L\left(x_{\sigma_b}^{[1]}\right) > L\left(x_{\sigma_g}^{[1]}\right)\right] > \frac{1}{2},$$

that is, by equation (3.18) we have $\mathcal{P}\left[\hat{I}(\sigma_b) > \hat{I}(\sigma_g)\right] > \frac{1}{2}$, which implies

$$\mathcal{P}\left[\hat{I}(\sigma_b) > \hat{I}(\sigma_g)\right] > \mathcal{P}\left[\hat{I}(\sigma_b) \leq \hat{I}(\sigma_g)\right]. \tag{3.21}$$

As before, let $\sigma^*$ denote the unique global optimum. Let $\eta \in \Sigma_0$, and let $\xi_0, \xi_1, ..., \xi_{n-1}, \xi_n$ be the shortest path from $\sigma^*$ to $\eta$, where $n = \kappa(\sigma^*, \eta)$. Take any $i$ such that $1 \leq i \leq n - 1$. We need to consider three cases here:

1. If $\sigma^* \in \xi_{i-1}$ and $\sigma^* \in \xi_{i+1}$ then $\xi_{i-1} \in \Sigma_g$ and $X \setminus \xi_i \in \Sigma_b$. We also note that according to Definition 6, $\xi_{i+1} = b(\xi_i)$ and $\xi_{i-1} \in \mathcal{H}(\xi_i)$. Thus, according to equation (3.5) and the inequality (3.21) derived above:

$$
\begin{aligned}
P(\xi_i, \xi_{i-1}) &= \sum_{i=1}^{|\mathcal{H}(\xi_i)|+1} \mathcal{P}[\xi_{i-1} \in B_2 : |B_2| = i] \cdot \mathcal{P}[|B_2| = i]/i \\
&= \sum_{i=1}^{|\mathcal{H}(\xi_i)|+1} \mathcal{P}[\hat{I}(\xi_{i-1}) \leq \hat{I}(\eta), \forall \eta \in \mathcal{H}(\xi_i) : |B_2| = i] \cdot \mathcal{P}[|B_2| = i]/i \\
&> \sum_{i=1}^{|\mathcal{H}(\xi_i)|+1} \mathcal{P}[\hat{I}(X \setminus \xi_i) \leq \hat{I}(\eta), \forall \eta \in \mathcal{H}(\xi_i) : |B_2| = i] \cdot \mathcal{P}[|B_2| = i]/i \\
&= \sum_{i=1}^{|\mathcal{H}(\xi_i)|+1} \mathcal{P}[\xi_{i+1} \in B_2 : |B_2| = i] \cdot \mathcal{P}[|B_2| = i]/i \\
&= P(\xi_i, \xi_{i+1}).
\end{aligned}
$$

2. If $\sigma^* \in \xi_{i-1}$ and $\sigma^* \notin \xi_{i+1}$ then $\xi_{i-1} \in \Sigma_g$ and $\xi_{i+1} \in \Sigma_b$. Furthermore, $\xi_i = b(\xi_{i-1}) = b(\xi_{i=1})$. Similarly to the first case we can use (3.5), or (3.3) if $\xi_i = X$, to show that $P(\xi_i, \xi_{i-1}) > P(\xi_i, \xi_{i+1})$.
3. If $\sigma^* \notin \xi_{i-1}$ and $\sigma^* \notin \xi_{i+1}$ then $X \setminus \xi_{i-1} \in \Sigma_g$ and $\xi_{i+1} \in \Sigma_b$. Furthermore, $\xi_{i-1} = b(\xi_i)$ and $\xi_{i+1} \in \mathcal{H}(\xi_i)$. Again, equation (3.5) can be used to show that $P(\xi_i, \xi_{i-1}) > P(\xi_i, \xi_{i+1})$.

Thus, we have that $P(\xi_i, \xi_{i-1}) > P(\xi_i, \xi_{i+1})$, and since this holds for any $i$ then it is clear that

$$P(\xi_{n-1}, \xi_{n-2}) \cdot \ldots \cdot P(\xi_1, \xi_0) > P(\xi_1, \xi_2) \cdot \ldots \cdot P(\xi_{n-1}, \xi_n). \tag{3.22}$$

Now we note that using a similar argument as above and noting that $\xi_n, X \setminus \xi_0 \in \Sigma_b$, whereas $\xi_0, X \setminus \xi_n \in \Sigma_g$, equations (3.21) and (3.6) imply that $P(\xi_n, \xi_{n-1}) > P(\xi_0, \xi_1)$, and putting this together with (3.22) we have

$$P^{\kappa(\eta, \sigma^*)}(\eta, \sigma^*) > P^{\kappa(\sigma^*, \eta)}(\sigma^*, \eta). \tag{3.23}$$

Now finally, to verify that equation (3.9) of Assumption 2 holds, let $\xi \in \mathcal{S}_0$. Then by definition $P^{\kappa(\eta, \xi)}(\eta, \xi) \geq P^{\kappa(\xi, \eta)}(\xi, \eta), \forall \eta \in \Sigma_0$. If $\xi \neq \sigma^*$ then we can select $\eta = \sigma^*$ and obtain $P^{\kappa(\sigma^*,\xi)}(\sigma^*,\xi) \geq P^{\kappa(\xi,\sigma^*)}(\xi,\sigma^*)$, which contradicts equation (3.23) above. Thus, we must have $\xi = \sigma^*$, which implies that $\mathcal{S}_0 = \mathcal{S}$ so Assumption 23.2is satisfied and global convergence follows by Theorem 3.4 above.

The assumption of Theorem 3.7 illustrates sufficient global convergence conditions on the partitioning and sampling on the one hand, and the estimation on the other. In particular, $E[Y(\sigma_g, \sigma_b)]$ depends only on the partitioning and sampling and the inequality (3.20), which is implicit in the assumption, can be satisfied by partitioning such that there is a certain amount of separation, or partial non-overlap, between the good and bad sets, and then by using enough sample points. On the other hand, given a fixed value of $E[Y(\sigma_g, \sigma_b)]$, the inequality (3.19) depends on the sample performance being sufficiently accurate and its ability to be satisfied by increasing the simulation time. However, verifying how much sample effort is 'enough' and when the accuracy of the sample performance is 'sufficient' may still be fairly difficult.

## 3.2 Selecting the Correct Move

As we noted in the introduction the main implementation issue due to the noisy performance is the fact that it now becomes more difficult to select which region should become the next most promising region, that is, to make the correct move on the state space fixed by the partitioning. The reason for this is that the performance of each region is now not only measured based on a randomly generated sample of solutions from the region, but the performance of each of these generated solutions is not accurately known.

A key observation in understanding how the NP method is still effective in the presence of noise is that for the method to proceed effectively, it is not necessary to obtain accurate performance estimates for each of the generated solutions - it suffices for this estimate to be good enough so that the correct region is selected. The essential issue is therefore the preservation of rank, which is sometimes referred to as ordinal optimization. We start by briefly discussing this area before continuing with methods for rigorously determining how much effort is needed in each region to preserve rank.

### 3.2.1 Ordinal Optimization

We start by defining the states (subsets of solutions) that should be selected by the NP method as it moves through the state space $\Sigma$ of all valid regions.

**Definition 3.8.** *We let $\Sigma_g \subset \Sigma$ denote all the subregions that contain the global optimum, that is, $\sigma \in \Sigma_g$ if and only if $x^* \in \sigma$. We refer to these subregions as the good subregions.*

Now recall that given a valid region $\sigma \in \Sigma$ and a set of sample solutions $\mathcal{D}_\sigma \subseteq \sigma$ that have been generated from the region, the performance of the region is estimated using the best performance among the sample solutions:

$$\hat{I}(\sigma) = \min_{x \in \mathcal{D}_\sigma} \hat{f}(x), \tag{3.24}$$

where $\hat{f}(x)$ is the estimated performance of $x \in \mathcal{D}_\sigma$. Unfortunately an accurate estimate of the performance is often computationally expensive. For example, a long simulation run must be made in the case of a simulation optimization problem, or a large number of data points must be used in the case of a data mining problem. Convergence to the true performance is also often slow. For example, if the performance is estimated using simulation, it is well known that the estimate $\hat{f}(x)$ converges to $f(x)$ at a rate that is at the most $O(\frac{1}{\sqrt{t}})$ in the simulation time $t$.

This is where it is helpful in practice that the NP method only needs to preserve rank, that is, it focuses on the ordinal rather than the cardinal values of the performance. In our notation, if it is desirable to move into a good region $\sigma_g \in \Sigma_g$ and this is being compared to another region $\sigma_b \in \Sigma \setminus \Sigma_g$, then it is sufficient that

$$\hat{I}(\sigma_g) < \hat{I}(\sigma_b), \tag{3.25}$$

that is, if the rank is preserved then the correct valid region will be selected. The advantage of this sufficing is that results from ordinal optimization show that the estimated rank of a random variable may converge to its true rank at an exponential rate even if the cardinal values converge at a much slower rate. The implication is that it is not necessary to accurately estimate $f(x)$ to obtain a sufficiently good estimate of the promising index.

The paradigm of ordinal optimization is based on two basic ideas (Ho, Sreenivas and Vakili, 1992; Ho, 1994; Dai, 1996; Lee, Lau and Ho, 1999). The first is that estimates of ordinal values may converge at an exponential rate even though the estimated cardinal values do not. As illustrated above, the NP method is always ordinal in this sense and we will see that this observation is therefore beneficial for any implementation of the algorithm. The second idea is that relaxing the goal of finding the optimal solution to finding a 'good enough' solution may also result in an exponential convergence rate. Adopting this perspective, there is a set $\mathcal{G} \subseteq X$ of feasible solutions that are of sufficient quality. These solutions are such that if $X \in \mathcal{G}$ then $f(x) \leq f(y)$ for all $x \in \mathcal{G}, y \notin \mathcal{G}$, that is, these are the $g = |\mathcal{G}|$ solutions that have the desired performance. To select a $x \in \mathcal{G}$ maintain a set $\mathcal{S}(k)$ of the $g$ best designs found by the $k$th iteration. The goal is to have at least one common element in the sets $\mathcal{G}$ and $\mathcal{S}(k)$. This may be considered goal softening from the original formulation requiring the global optimum. To see that this softened goal is rapidly achieved consider the probability of misalignment $Q(k) = 1 - P\big[|\mathcal{G} \cap \mathcal{S}(k)| \geq 1\big]$. This is the probability that there is no good design in the set $\mathcal{S}(k)$. It is shown that, given certain mild conditions, this misalignment probability converges to zero at an exponential rate if the feasible region is sampled uniformly.

### 3.2.2 Ranking and Selection

We now consider how the ideas of ordinal optimization, namely goal softening and ordinal comparison, can be used to guide an efficient implementation of the NP method for solving problems with noisy performance. As the method selects a new most promising region in each iteration, this selection can be considered a success if the selected region contains the true global optimum. We explored in Chapter 2 how the efficiency of the method depends on the ability to make this selection correctly with a high probability, and since the selection is more difficult with noisy performance it would be of practical interest if a minimum probability of success could be guaranteed in each iteration.

As previously noted, when applying the NP method to a problem with noisy performance there are two sources of randomness that complicate the selection of the correct subregion. First, there is a sampling error due to a relatively small sample of feasible solutions being used to estimate the performance of an often large set. Second, there is noise in the performance estimate of each of those samples. It is important to observe that the former of these elements implies that the variation within a subregion depends on the size of the region among other factors. As an extreme case consider a singleton region that is being compared to the entire complimentary region. (That is, a region containing only one solution being compared to a region containing all of the other solutions.) Clearly the first source of randomness has been completely eliminated in the singleton region, whereas it may account for almost all of the randomness in the complimentary region. This implies that to make better use of the computational effort the number of sample solutions generated from each region should be variable and dependent on the variation within the region.

### Two-Stage Sampling

The above discussions identify two potential shortcomings of the pure NP method: the success probability in each iteration is not guaranteed and there may be considerable waste involved in how the sample solutions are generated. We now show how this can be addressed by incorporating a statistical selection method into the NP framework in order to compare the subregions (Kim and Nelson, 2006). In particular, we will show how to use the classic Rinott's two-stage ranking and selection procedure for selecting the best subregion.

To state this approach rigorously let $\mathcal{D}_{ij}(k)$ be the $i$th set of sample points selected from the region $\sigma_j(k)$ using a uniform random sampling procedure to generate feasible solutions from each region, $i \geq 1$, $j = 1, 2, ..., M + 1$ in the $k$th iteration. As before, let $N = |\mathcal{D}_{ij}(k)|$ denote the number of sample solutions generated and assume that it is constant. As usual let $x \in \mathcal{D}_{ij}(k)$

denote a point in that set and let $\hat{f}(x)$ be an estimate of the performance of this solution. Then in the $k$th iteration,

$$\hat{Y}_{ij}(k) = \min_{x \in \mathcal{D}_{ij}(k)} \hat{f}(x)$$

is an estimate of the performance of the region $\sigma_j$, which can also be referred to as the $i$th system performance for the $j$th system, $i \geq 1$, $j = 1, 2, ..., M+1$. The two-stage ranking and selection procedure first obtains $n_0$ such system estimates, and then uses that information to determine the total number $N_j$ of system estimates needed from the $j$th system, that is, subregion $\sigma_j(k)$.

More precisely, the Two-Stage NP (TSNP) procedure is as follows:

## Algorithm *TSNP*

### Partitioning

Step 1.  Given a current most promising region $\sigma(k)$, partition $\sigma(k)$ into $M$ subregions $\sigma_1(k), ..., \sigma_M(k)$ and aggregate the complimentary region $X \setminus \sigma(k)$ into one region $\sigma_{M+1}(k)$.

### Stage I Sampling

Step 2.  Let $i = 1$.

Step 3.  Use uniform sampling to obtain a set $\mathcal{D}_{ij}(k)$ of $N$ sample points from region $j = 1, 2, ..., M+1$.

Step 4.  Use discrete event simulation of the system to obtain a sample performance $\hat{f}(x)$ for every $x \in \mathcal{D}_{ij}(k)$ and estimate the performance of the region as

$$\hat{Y}_{ij}(k) = \min_{x \in \mathcal{D}_{ij}(k)} \hat{f}(x), \tag{3.26}$$

$j = 1, 2, ..., M+1$.

Step 5.  If $i = n_0$ continue to Step 6. Otherwise let $i = i + 1$ and go back to Step 3.

### Stage II Sampling

Step 6.  Calculate the first-stage sample means and variance

$$\bar{Y}_j^{(1)}(k) = \frac{1}{n_0} \sum_{i=1}^{n_0} \hat{Y}_{ij}(k), \tag{3.27}$$

and

$$S_j^2(k) = \frac{1}{n_0 - 1} \sum_{i=1}^{n_0} \left[ \hat{Y}_{ij}(k) - \bar{Y}_j^{(1)}(k) \right]^2, \tag{3.28}$$

for $j = 1, 2, ..., M+1$.

Step 7.  Compute the total sample size

$$N_j \left( n_0, M, P^*, \bar{Y}_j^{(1)}(k), S_j^2(k) \right) \tag{3.29}$$

using the desired ranking-and-selection procedures (see below).

Step 8.   Obtain

$$\left(n_0, M, P^*, \bar{X}_j^{(1)}(k), S_j^2(k)\right) - n_0$$

more estimates of the subregion performance as in Step 2 - Step 5
above, that is $(N_j(k) - n_0) \cdot N$ more sample points.

**Estimating promising index Values**

Step 9.   Let the overall sample mean be the promising index for each region,

$$\hat{I}\left(\sigma_j(k)\right) = \bar{Y}_j(k) = \frac{1}{N_j(k)} \sum_{i=1}^{N_j(k)} \hat{Y}_{ij}(k), \qquad (3.30)$$

$j = 1, 2, ..., M + 1.$

**Determining the Next Move**

Step 10.  Calculate the next most promising region $\sigma(k+1)$ as in the Pure
NP Algorithm.

Step 11.  Update the counters $\{\mathcal{N}_k(\sigma)\}_{\sigma \in \Sigma}$ and $\hat{\sigma}^*(k+1)$ as in Step 5 of
Algorithm NP.

Step 12.  If the stopping rule is not satisfied let $k = k + 1$ and go back to
Step 1.

Note that for $n_0 = 1$, with Steps 6-8 omitted, and $\hat{I}(\sigma_i(k)) = \hat{Y}_{1i}$ replacing
equation (3.30), this new algorithm reduces to the pure NP algorithm. On the
other hand, by selecting $M = |X|$, the algorithm reduces to a pure two-stage
ranking-and-selection procedure.

## Other Statistical Selection Procedures

Any statistical selection procedure that guarantees correct selection within
an indifference zone with a given probability can be incorporated into the NP
framework and the TSNP algorithm as shown above. Numerous such methods
have been proposed in the literature and from an implementation standpoint
the only difference is how the total sample size (3.29) is obtained.

The best-known classical methods for ranking and selection is proba-
bly Rinott's procedure. Given the indifference zone $\delta$ and a constant $h = h(n_0, M, P^*)$ that can be obtained by solving an integral numerically, the
amount of sample effort is determined as

$$N_j\left(n_0, M, P^*, \bar{X}_j^{(1)}(k), S_j^2(k)\right) = \max\left\{n_0, \left\lceil \frac{h^2 S_j^2(k)}{\varepsilon} \right\rceil\right\}. \qquad (3.31)$$

We refer to the TSNP algorithm using (3.31) for equation (3.29) as the
NP/Rinott algorithm.

As a statistical selection procedure the Rinott approach has some well-
documented limitations. From the perspective of TSNP, the following are the
primary issues:

- The systems compared are assumed to be independent, that is, for NP/Rinott samples must be obtained independently for each region in both Step 3 and Step 8 of the algorithm.
- The Rinott procedure is based on the assumption of least-favorable configuration between the systems. This is not likely in practice and makes the procedure quite conservative in the sense that (3.31) tends to prescribe more sample effort than is really needed to assure the probability of correct selection.
- Finally, note that (3.31) does not depend on the mean performance, only the estimated variance. Thus, large amounts of sample efforts may be prescribed to regions where the mean indicates very poor performance.

The remainder of this section discusses two methods that can be used to alleviate these three issues.

One relatively simple way in which some of the shortcomings of the Rinott procedure can be addressed is to add filtering or subset selection. According to this approach first calculate the quantity

$$W_{ij} = t \left( \frac{S_i^2 + S_j^2}{n_0} \right)^{1/2} \tag{3.32}$$

for all $i \neq j$, and then use

$$N_j \left( n_0, M, P^*, \bar{X}_j^{(1)}(k), S_j^2(k) \right)$$

$$= \begin{cases} \max \left\{ n_0, \left\lceil \frac{h^2 S_j^2(k)}{\varepsilon} \right\rceil \right\} & \text{if } \bar{X}_i \leq \bar{X}_j + (W_{ij} - \varepsilon)^+, \forall i \neq j, \\ n_0, & \text{otherwise.} \end{cases} \tag{3.33}$$

The TSNP algorithm is referred as the NP/Subset/Rinott algorithm by using (3.33) for equation (3.29).

When comparing simulated systems it is well known that it is beneficial to use common random numbers and thus make the systems dependent. Hence it is of interest to consider statistical selection methods that allow for correlated systems. One such method is proposed by Matejcik and Nelson (1995) and this can also be incorporated into the NP framework. Here the total sample size is calculated according to

$$N_j \left( n_0, M, P^*, \bar{Y}_j^{(1)}(k), S^2 \right) = \max \left\{ n_0, \left\lceil \left( \frac{gS}{\varepsilon} \right)^2 \right\rceil \right\}, \tag{3.34}$$

where the variance estimate must now account for possible interaction, that is, equation (3.28) is replaced with

$$S^2 = \frac{2 \sum_{j=1}^{M+1} \sum_{i=1}^{n_0} \left( \hat{Y}_{ij} - \bar{Y}_{i.} - \bar{Y}_{.j} + \bar{Y}_{..} \right)^2}{(M)(n_0 - 1)}. \tag{3.35}$$

Furthermore, $g = T^{\alpha}_{M,(M)(n_0-1),0.5}$ is an equicoordinate critical point of the equicorrelated multivariate central $t$- distribution. This constant can be found in Hochberg and Tamhane (1987), Appendix 3, Table; Bechhofer(1995); or by using the FORTRAN program AS251 of Dunnet (1989).

## 3.3 Time Until Convergence

It was shown in the last chapter that the TSNP algorithm converges asymptotically. In this section we consider how quickly it converges. By Definition 3.1, for the algorithm to correctly consider the optimum $\sigma^*$ as the best solution, this state must be visited at least once. Hence it is of interest to look at the expected time until the algorithm visits this state for the first time. We would like this to be as short as possible, and the next theorem provides an upper bound for this expected time.

As stated earlier both Algorithm NP and Algorithm TSNP generate a Markov chain and the stationary distribution of this chain can be used for inference about the convergence of the algorithm. To state this precisely, we need the following technical assumption, which can be made without loss of generality.

**Assumption 3.3** *Assume that* $\forall \eta \in \Sigma, \exists X \in \eta, \xi \in X \setminus \eta$, *such that* $P[L(X) < L(\xi)] < 1$.

With this assumption the following proposition follows:

**Proposition 3.9.** *If Assumption 3.3 holds then Algorithm TSNP generates an irreducible recurrent Markov chain such that its unique stationary distribution* $\pi$ *satisfies*

$$\lim_{k \to \infty} \pi\left(\hat{\sigma}^*(k)\right) > \pi(\eta), \ \forall \eta \in \Sigma_0 \setminus \{\sigma^*\}, \ w.p.1. \tag{3.36}$$

*In other words, the algorithm converges to a maximum of the stationary distribution over all singleton regions.*

*Proof:* The proof of this proposition is similar to that to Theorem 3.2, which holds for a slightly more general situation. Recall that it is clear that $\{\sigma(k)\}_{k=0}^{\infty}$ is a Markov chain, and that it is irreducible by Assumption 3.3. Since $\Sigma$ is finite, the Markov chain is positive recurrent with a unique stationary distribution. Furthermore, it is well known (see e.g. Ross 1996) that $\lim_{k \to \infty} \frac{N_k(\eta)}{k} = \pi(\eta)$, which implies that, in the limit, the most frequently visited region maximizes the stationary distribution. Since $\hat{\sigma}^*(k) = \arg\max_{\sigma \in \Sigma_0} N_k(\sigma)$ the proposition follows.

To state the main convergence theorem we need the usual assumption of ranking and selection methods (Bechhofer, Santner and Goldsman 1995), namely that the observations are normally distributed.

**Assumption 3.4** *Assume that $X_{ij} \sim \mathcal{N}(\mu_j, \nu_j^2)$, is normally distributed with mean $\mu_j$ and variance $\nu_j^2$ for all $j \in \{1, 2, ..., M+1\}$, and $i \in \{1, 2, ..., N_j(k)\}$, $k \geq 1$.*

In practice this is not likely to hold exactly, but our procedure, as with ranking-and-selection procedures in general, may be robust with respect to deviations from this assumption. In addition to the normality assumption, we need to be able to distinguish between the optimum and other solutions.

**Assumption 3.5** *Assume that the indifference zone $\epsilon$ satisfies*

$$\epsilon \leq \min_{X \in X \setminus X_{opt}} f(X) - f(X_{opt}).$$

In practice $f(X) - f(X_{opt})$ is unknown and we simply select $\varepsilon$ based on what performance difference we are indifferent about.

**Theorem 3.10.** *Let Assumption 3.3 - 3.5 hold and assume that $P^* > 0.5$. Let $T_1$ denote the first time Algorithm TSNP visits the optimal solution. Then*

$$E[T_1] \leq \frac{d^*}{2P^* - 1}. \tag{3.37}$$

*Proof:* Recall that the Markov chain $\{\sigma(k)\}_{k=1}^{\infty}$ has a minimum success probability of $P^*$ given its current state $\sigma(k) \in \Sigma$, that is, with probability of at least $P^*$, $\sigma(k+1)$ will be closer to $\sigma^*$ than $\sigma(k)$ in terms of the number of transitions required to move between the regions. Now imagine a Markov chain that is identical to $\{\sigma(k)\}$ except that this success probability is even and equal to $P^*$ for every state $\sigma \in \Sigma$. Now note that since the success probability is constant, the exact state is not of any consequence, but rather the number of transitions it takes to move from the current state $\sigma(k)$ to the optimum. The maximum such distance is $2d^*$, and we can therefore, without losing any information, reduce the state space to $\mathcal{S} = \{0, 1, 2, ...., 2d^*\}$. With this representation the entire feasible region $X$ corresponds to state $d^*$, and we can let the global optimum correspond to state zero. Given a state $x \in \mathcal{S}$ the probability of moving to $x-1$ is fixed and equal to $P^*$, and the probability of moving to $x+1$ is equal to $1 - P^*$, regardless of the state. Therefore, the new Markov chain is a simple random walk. Furthermore, it is clear that $E[T_1] \leq E[T_1']$, where $T_1'$ is the first time the random walk visits $\sigma^*$ if it starts in state $d^*$, which corresponds to $X$, the starting state of Algorithm TSNP.

Hence, if we calculate the expected hitting time for the random walk this automatically gives us an upper bound for the original Markov chain. Furthermore, since we are only interested in the time the global optimum is found for the first time, we can assume 0 is an absorbing barrier and look at the time of absorption. Note also that $2d^*$ is a reflective barrier. Then it is known that the expected time $T$ of absorption when starting in state $u$ is (Weesakul, 1961)

$$E_u[T] = \frac{u}{2P^* - 1} + \frac{(1 - P^*)^{2d^* + 1}}{(P^*)^{2d^*}(2P^* - 1)^2}\left(1 - \left(\frac{P^*}{1 - P^*}\right)^u\right). \tag{3.38}$$

Thus, for $u = d^*$, that is, when the algorithm starts in state $\sigma(0) = X$, we have

$$E_{d^*}[T] = \frac{d^*}{2P^* - 1} + \frac{(1 - P^*)^{2d^* + 1}}{(P^*)^{2d^*}(2P^* - 1)^2}\left(1 - \left(\frac{P^*}{1 - P^*}\right)^{d^*}\right) \equiv C_1. \tag{3.39}$$

Now since $P* > 0.5$ then $\frac{P^*}{1 - P*} > 1$ so

$$\left(1 - \left(\frac{P^*}{1 - P^*}\right)^{d^*}\right) < 0,$$

and hence

$$\frac{(1 - P^*)^{2d^* + 1}}{(P^*)^{2d^*}(2P^* - 1)^2}\left(1 - \left(\frac{P^*}{1 - P^*}\right)^{d^*}\right) < 0.$$

Therefore,

$$E[T_1] \le E_{d^*}[T] \le \frac{d^*}{2P^* - 1},$$

which proves the theorem.

From the proof above it is clear that a tighter bound on $E[T_1]$ can be obtained by using $C_1$ as defined by equation (6.12) above. However, unless both $P^*$ is very close to 0.5 and $d^*$ is small, then the difference in the two bounds will be negligible. Since the value of $P^*$ is given as an input parameter, a selection of, say, $P^* \ge 0.51$ will always ensure an adequate bound and there is therefore little practical benefit from using the tighter bound. In addition to the first time the optimum is found, the time that elapses until the optimum is visited again is also of interest.

**Theorem 3.11.** *Let Assumptions 1-3 hold and assume that $P^* > 0.5$. Let $T_2$ denote the time between the first and second time Algorithm TSNP visits the optimal solution. Then*

$$E[T_2] \le \frac{1 - P^*(1 - 2P^*)}{2P^* - 1}. \tag{3.40}$$

*Proof:* As before let $T$ denote the absorption time of the random walk with a reflective and absorbing barrier defined by a constant success probability $P^*$. Since in the first transition after visiting $\sigma^*$ the Markov chain either stays at $\sigma^*$, or moves to $s(\sigma^*)$, it is clear that

$$E[T_2] \le P^* \cdot 1 + (1 - P^*)E_1[T].$$

**Table 3.1.** Expected first hitting time of the optimum.

| | Maximum Depth $(d^*)$ | | | | |
|---|---|---|---|---|---|
| Success Prob. | 2 | 5 | 10 | 20 | 30 |
| 55% | 20 | 50 | 100 | 200 | 300 |
| 60% | 10 | 25 | 50 | 100 | 150 |
| 65% | 7 | 17 | 33 | 67 | 100 |
| 70% | 5 | 13 | 25 | 50 | 75 |
| 75% | 4 | 10 | 20 | 40 | 60 |
| 80% | 3 | 8 | 17 | 33 | 50 |
| 85% | 3 | 7 | 14 | 29 | 43 |
| 90% | 3 | 6 | 13 | 25 | 38 |
| 95% | 2 | 6 | 11 | 22 | 33 |

Letting $u = 1$ in equation (3.38) gives

$$E_1[T] = \frac{1}{2P^* - 1} + \frac{(1 - P^*)^{2d^*+1}}{(P^*)^{2d^*}(2P^* - 1)^2}\left(1 - \left(\frac{P^*}{1 - P^*}\right)^1\right) \leq \frac{1}{2P^* - 1},$$

so

$$E[T_2] \leq P^* + \frac{1}{2P^* - 1},$$

which proves the theorem.

Similarly as with the bounds on $E[T_1]$ the bounds on $T_1$ can be tightened by using

$$C_2 \equiv P^* + (1 - P^*) \cdot \left(\frac{1}{2P^* - 1} + \frac{(1 - P^*)^{2d^*+1}}{(P^*)^{2d^*}(2P^* - 1)^2}\left(1 - \left(\frac{P^*}{1 - P^*}\right)^1\right)\right).$$
$$(3.41)$$

However, as before, unless both $P^*$ is very close to 0.5 and $d^*$ is small the difference will be very small and there is therefore no practical benefit from using the more complicated but tighter bound.

Now lets consider whether an optimal selection probability $P^*(n_0, M)$ and can be found. It is clear that as $P^*(n_0, M)$ increases, $E[T_1]$, that is, the expected time until the global optimum is encountered, decreases. This occurs, however, at a decreasing rate. On the other hand, as $P^*(n_0, M)$ increases, $h(n_0, M, P^*)$ also increases and this occurs at an increasing rate. Therefore, an optimal probability is somewhere between the extreme values of $P^*(n_0, M) = 0.5$ and $P^*(n_0, M) = 1$. However, since the second-stage sample size depends on the sample variance from the first stage sampling and the indifference zone, both of which are problem dependent and unknown, the same holds for the optimal value of $P^*(n_0, M)$. It is therefore not possible to give an *a priori* prescription for the optimal probability. Nonetheless, more useful information can be extracted from our random walk analysis.

Another quantity of interest when applying the Algorithm TSNP is the probability of the first maximum depth region visited being the one corresponding to the global optimum. If this probability is sufficiently high then a reasonable stopping rule would be to stop whenever maximum depth is reached. We can again use a random walk analysis, this time for a simple random walk with two absorbing barriers, to calculate this probability.

**Theorem 3.12.** *Let Assumption 3.3-3.5 hold and assume that $P^* > 0.5$. Let $\hat{\sigma}$ denote the first maximum depth region visited. Then*

$$\psi \equiv P\left[\hat{\sigma} = \sigma^*_{opt}\right] = \frac{(P^*)^{d^*}}{(1 - P^*)^{d^*} + (P^*)^{d^*}}. \tag{3.42}$$

*Proof:* Since the success probability is constant we can again consider the random walk with state space $\mathcal{S} = \{0, 1, ..., 2d^*\}$ defined in the proof of Theorem 3 above. Here the only question is thus whether state 0 or $2d^*$ will be visited first; that is, the probability that the first maximum depth visited contains the global optimum is equal to the probability that the random walk visits state 0 before it visits state $2d^*$. This probability is thus equal to the absorption probability at zero for a simple random walk with two absorbing barriers, which can for example be found on p. 32 in Cox and Miller (1965) for a random walk $\{X_n\}$ with upward probability $p = P[X_{n+1} = X_n + 1]$ and downward probability $q = P[X_{n+1} = X_n - 1]$, and absorbing barriers at $a$ and $-b$:

$$P\left[X_N = -b\right] = \begin{cases} q^b \frac{p^a - q^a}{p^{a+b} - q^{a+b}} & p \neq q, \\ \frac{a}{a+b} & p = q, \end{cases}$$

where $N$ is the absorption time. Here we have $q = P^* > 0.5$ and $p = 1 - q$ so $q \neq p$, and $a = b = d^*$. Thus, the expression simplifies to equation (3.42) of the theorem. We now obtain the following stopping rule:

Stop if $d(\sigma(k)) = d^*$, and report the final solution $\sigma(k)$ with the probability

$$P\left[\hat{\sigma} = \sigma^*_{opt}\right] = \frac{(P^*)^{d^*}}{(1 - P^*)^{d^*} + (P^*)^{d^*}}$$

that the performance of this solution is within indifference zone $\varepsilon$ of the optimal performance.

Otherwise let $k = k + 1$ and go back to Step 1.

Note that this stopping rule assumes that $P^* > \frac{1}{2}$, which is a necessary condition for convergence. We also note that it is possible to use the same random walk analysis to calculate how many iterations are to be expected before the stopping criterion developed above is satisfied, that is, the expected time until maximum depth is reached for the first time.

We conclude this section with a few comments on the practical implications of the algorithm. We made the assumption that the indifference zone $\varepsilon$ is selected such that it differentiates between the best and second-best solution (see Assumption 4.3). In practice this is not likely to be true, in which case the algorithm would not be assured to converge to the optimal solution, but rather to one of the solutions within the indifference zone of the optimal performance. This, of course, is consistent with our understanding of what an indifference zone should be. Along with $\varepsilon$, we can now also choose a target probability $\psi$ by which we wish to correctly terminate. Furthermore, we can select $d^*$ so that we terminate the search at a set of a desired size. That is, instead of terminating at a singleton we might select a smaller $d^*$ value that implies that we terminate at a set that reduces the feasible region by an arbitrary amount. For example, we can choose $d^*$ such that the TSNP algorithm reduces the search to a set of 30-50 solutions, which could then be followed up with a pure ranking-and-selection procedure for determining the best solution. Thus, we can *a priori* set reasonable or satisfactory goals in terms of $\varepsilon$, $d^*$, and $\psi$, and then terminate the search the first time maximum depth is reached. Thus, before starting the algorithm we go through the following initialization steps:

### Initialization

1. Determine the size of the desired final set and calculate the corresponding $d^*$ value.
2. Determine a desired indifference zone $\varepsilon > 0$ and probability $\psi$ of correct selection.
3. Calculate $P^*$ according to equation (3.42).

As the type of problems addressed in this book are extremely difficult to solve and an optimal solution is typically an unrealistic goal, this type of goal softening is of considerable importance.

# 4

# Mathematical Programming in the NP Framework

Mathematical programming methods have been effectively used to solve a wider range of problems that contain sufficient structure to guide the search. In this chapter we are primarily interested in problems that can be stated as *mixed integer programs* (MIP). For such problems there may be one set of discrete variables and one set of continues variables and the objective function and constraints are both linear. A general MIP can be stated mathematically as follows:

$$z_{MIP} = \min_{x,y \in X} c^1 x + c^2 y, \qquad (4.1)$$

where $X = \left\{ x \in Z_+^n, y \in R^n : A^1 x + A^2 y \leq b \right\}$ and we use $z_{MIP}$ to denote any linear objective function, that is, $z_{MIP} = f(x) = cx$. While some large-scale MIPs can be solved efficiently using exact mathematical programming methods (Atamturk and Savelsbergh, 2005), complex applications often give rise to MIPs where exact solutions can only be found for relatively small problems. As before, we are interested in complex large-scale problems where traditional exact methods are not effective and the NP method has been proven to be very useful. However, even in such cases it may be possible to take advantage of exact mathematical programming methods by incorporating them into the NP framework. The NP method therefore provides a framework for combining the complimentary benefits of two optimization approaches that have traditionally been studied separately, namely mathematical programming and metaheuristics.

In this chapter we discuss how in the NP method mathematical programming methods can be utilized to find intelligent partitioning, generate good feasible sample solutions, and define an improved promising index. We will restrict our attention primarily to problems that can be formulated as MIPs. The difficulty of these problems hence arises from the large-scale nature of the problems and the fact that at least some of the decision variables are discrete. We do this to highlight the connection and synergy between the NP method and traditional mathematical programming methods for MIPs. However, the application of the NP method is certainly not limited to problems with linear objective functions and constraints and in the final section of the chapter we briefly discuss extending the results of this chapter to non-linear problems. Before addressing how to incorporate mathematical programming into the NP method we need to briefly review some relevant mathematical programming concepts. For more information the reader can consult Nemhauser and Wolsey (1988), Wolsey (1998), and Aardal, Nembauser and Weismantel (2005).

## 4.1 Mathematical Programming

As was briefly reviewed in Chapter 1, there are the two primary classes of mathematical programming methods that can be used to solve discrete problems: branching methods and decomposition methods. We have already reviewed how for minimization problems branching methods focus on obtaining tight lower bounds $\underline{z} \leq z^*$ for each branch and then use these bounds to eliminate branches where the lower bound is worse than some known feasible solution $x^0 \in X$, that is, $cx^0 < \underline{z}$. Such branching can be done in the same way as partitioning in the NP method, but the NP method shifts the primary computational effort to generating feasible solution (upper bounds), which is often much more effective for complex problems.

In the following sections we briefly review two mathematical programming concepts: relaxations and column generation. These are chosen solely because they will be used in later sections as illustrative examples for how to incorporate mathematical programming into the NP framework. Since a comprehensive treatment of mathematical programming for discrete problems is outside the scope of this book many other important methods are not mentioned.

### 4.1.1 Relaxations

Relaxations play a key role in the use of mathematical programming for solving discrete optimization problems. The idea of a relaxation is to modify the constraints or the objective function in some way that makes the problem easier to solve and assures that the optimal solution to the relaxed problem is a lower bound on the original problem. Formally, we say that a problem

$$z_{RP} = \min_{x,y} \left\{ g(x,y) : x, y \in X^{(R)} \right\} \qquad (4.2)$$

is a *relaxation* of (4.1) if the following two conditions hold:

$$X = \left\{ x \in Z_+^n, y \in R^n : A^1 x + A^2 y \leq b \right\} \subseteq X^{(R)}, \qquad (4.3)$$

$$g(x,y) \leq c^1 x + c^2 y, \forall x, y \in X. \qquad (4.4)$$

Thus, the feasible region $X$ for the original is contained in the feasible region $X^{(R)}$ of the relaxed problem, and the objective function $g(x,y)$ of the relaxed problem is dominated by the objective function $c^1 x + c^2 y$ of the original.

It follows that $z_{RP} \leq z_{MIP}$, that is, the solution to the relaxed problem provides a bound on the original problem. However, the solution to the relaxed problem (4.2) is in general not feasible for the original problem (4.1) so solving relaxations does usually not generate feasible solutions.

Relaxations can be either based on generally applicable methods or on problem-specific methods. A simple general relaxation method for a MIP as defined by equation (4.1) is the *linear programming (LP) relaxation*:

$$z_{LP} = \min_{x,y} c^1 x + c^2 y$$
$$A^1 x + A^2 y \leq b$$
$$x \in R_+^n \tag{4.5}$$
$$y \in R^n.$$

The only difference is (4.5) where $x$ is now allowed to take any continuous value. With this relaxation, the problem becomes a linear program (LP) and methods exist to solve large-scale LPs very quickly. The LP relaxation is threfore much easier to solve than the original problem, and solving it provides a lower bound on the original MIP.

In previous chapters we have noted that it is common for practical problems to have one set of constraints that is hard and another that is easy. This is for example true for the TSP introduced in Chapter 2, where the assignment constraints are easy but the subtour elimination constraints are hard, and for the resource-constrained project scheduling problem of Chapter 1, where the precedence constraints are easy but the resource constraints are hard. We now consider relaxations for such problems. For notational simplicity, assume that we have a pure integer program (IP) with two sets of constraints.

$$z_{IP} = \min_x cx$$
$$A^1 x \leq b^1 \tag{4.6}$$
$$A^2 x \leq b^2 \tag{4.7}$$
$$x \in Z_+^n$$

The problem is complicated by one set of constraints, namely equation (4.7). A very simple relaxation would simply drop these constraints. For the TSP this would result in an easy to solve assignment problem and for resource-constrained project scheduling this would result in a project scheduling problem that can be solved quickly using the critical path method. Thus, lower bounds are easily obtained but such a relaxation is typically not very useful since the resulting bound is not very tight.

A more useful relaxation for problems with complicating constraints is the *Lagrangian relaxation* (LR) where the hard constraints are added to the objective function as shown in the following program.

$$z_{LR}(\lambda) = \min_x cx + \lambda \left( b^2 - A^2 x \right) \tag{4.8}$$

$$A^1 x \leq b^1$$
$$x \in Z_+^n$$

for some $\lambda \geq 0$. This is easily seen to be a relaxation since the original feasible region is clearly a subset of the new feasible region and for any $\lambda \geq 0$, $z_{LR}(\lambda) \leq z_{IP}$. Importantly, the LR problem is easy to solve since the complicating constraints are no longer present. Furthermore, it often results in a fairly tight and hence useful bound.

The quality of the LR bounds, that is, the gap $z_{IP} - z_{LR}(\lambda)$, clearly depends on the choice of $\lambda$, which is called the Lagrangian multiplier. The *Lagrangian dual* (LD)

$$z_{LD} = \max_{\lambda} z_{LR}(\lambda) \tag{4.9}$$

gives the tightest possible bounds for all Lagrangian relaxation problems. The Lagrangian dual can be seen to be a piece-wise linear optimization problem and it is traditionally solved using a subgradient algorithm, although other more efficient methods have been developed more recently (Frangioni, 2005).

We finally note that the quality of the LD bound can be seen to be at least as good as the LP relaxation bound, that is, for any (4.1) it is true that (Nemhauser and Wolsey, 1988)

$$z_{LP} \leq z_{LD} \leq z_{IP}. \tag{4.10}$$

However, the LD bound usually requires more computational effort.

There are numerous other relaxation methods and by definition they all provide a lower bound on the performance of the original problem (4.1). As we will see later in this chapter, by incorporating it into the NP method such a lower bound can be used to compliment the upper bound found by heuristically generating feasible sample solutions. The NP method hence naturally combines mathematical programming and heuristics in a single framework.

### 4.1.2 Column Generation

Lagrangian relaxation can be thought of as a decomposition method with respect to the rows (constraints) since it moves one or more rows into the objective function. The dual of decomposing with respect to rows is decomposition with respect to the columns (variables). One such method is the *Dantzig-Wolfe (DW)* reformulation (Vanderbeck and Savelsbergh, 2006). In fact, the DW reformulation can be seen to be the dual problem of the Lagrangian dual so in this case the duality between decomposing rows (constraint) and columns (variables) exists in a precise sense.

The usefulness of the DW decomposition for solving integer problem is that it divides the optimization problem into two parts: a master problem and subproblem(s). It is possible to reformulate any problem (4.1) as an appropriate master problem, but instead of solving the master problem directly, we usually solve what is called the *restricted master problem* (RMP). The only difference between the master problem and the RMP is that the RMP uses a (small) subset of the variables, which in this context are usually referred to as columns. Unless all of the columns (variables) that are not included in this subset are zero in an optimal solution solving the RMP does not yield an optimal solution to the original problem. The next step is therefore to solve subproblem(s) to determine if any of the remaining columns should be added. Specifically, the objective function of the subproblems is the reduced cost corresponding to a set of columns, and if there are negative reduced costs then the corresponding columns should be added and a new RMP solved. This iteration between the RMP and the subproblem(s) is repeated until there are no negative reduced costs, which indicates that the optimal solution has been found.

The DW decomposition approach of repeatedly solving a RMP and corresponding subproblem(s) starts with a small set of columns (variables) and iteratively generates additional columns until the optimal solution is found. In many large-scale problems this only requires explicitly considering a small fraction of all of the variables. Such a *column generation* approach is therefore particularly useful when there is huge number of variables that grows exponentially in the input parameters (Villeneuve et al., 2005). This is indeed the case for many of the kind of large-scale discrete optimization problems that we study in this book.

## 4.2 NP and Mathematical Programming

In Section 2 we presented the NP method as a metaheuristic, which is a natural interpretation due to its focus on generating feasible solutions. However, it is also closely related to certain mathematical programming methods. In this section we discuss its connections, similarities and uniqueness as it relates to two such methods: branch-and-bound and dynamic programming.

### 4.2.1 Branch-and-Bound

Recall that there are two main categories of methods used to solve discrete problems: branching methods and decomposition methods. In the previous section we discussed decomposition methods with respect to both the constraints (relaxations) and variables (DW decomposition) and we will see later how those can be incorporated into the NP framework. The result are hybrid algorithms that are more effective and efficient than the mathematical programming methods alone. In this section we develop further the previously made observation concerning similarities between the NP method and branching methods.

As previously stated, branching methods solve discrete optimization problems by dividing the feasible region into partitions called branches and then obtain lower bounds $\underline{z} \leq z^*$ on the performance of each branch. If for a particular branch the bound is such that it proves that all solutions in that branch are no better than some known feasible solution $x^0 \in X$, that is, $cx^0 < \underline{z}$, then this branch can be eliminated or fathomed. Branches that cannot be fathomed are branched further until eventually all of the feasible solutions are accounted for. The computational effort of branching methods usually focuses on obtaining tight lower bounds.

Branching in branch-and-bound and partitioning in the NP-method both generate partitions of the feasible region, that is, disjoint subsets covering the entire space. There is therefore a one-to-one correspondence between these aspects of the methods and they therefore impose the same type of structure on the feasible region. On the other hand, the manner in which the feasible region is searched given this structure is quite different.

Branch-and-bound and its many variants focus the computational effort on finding good lower bounds (Balas and Toth, 1995; Beale, 1979). For example, the branch-and-cut algorithm combines branch-and-bound with the generation of cutting planes that improve the formulation (Caprara and Fishetti, 1997; Martin, 2001; Padberg, 2005). The generation of cutting planes at each node is computationally expensive, but the improved formulation results in better relaxations and hence tighter bounds. On the contrary, the NP method focuses the computational effort on generating feasible solutions, which can be viewed as upper bounds, and uses these feasible solutions to calculate the promising index of the region. The promising index is very flexible and as discussed in Chapter 2 can both use the sampling information directly and incorporate exact or probabilistic lower bounds. The focus on sampling and the ability to incorporate a variety of domain knowledge and heuristics makes the NP method more applicable for large-scale problems and for problems that are too complex for the development of tight lower bounds.

### 4.2.2 Dynamic Programming

Dynamic programming is an often efficient exact approach for solving a class of discrete optimization problems that can be formulated in the following manner (Bertsekas, 2000).

$$z = \min_{x_1, \dots, x_T} \sum_{t=1}^{T} g_t\left(s_{t-1}, x_t\right), \tag{4.11}$$

$$s_t = \phi_t\left(s_{t-1}, x_t\right), t = 1, \dots, T - 1.$$

and $s_0$ is given. The variable $s_t$ is called the state at time $t = 1, \dots, T$; and each $t$ is referred to as the time period or stage. The particular definition of the states and time stages depend on the application.

Following a standard DP approach, we will develop a recursive relationship to solve this problem. To that end, define

$$z_k\left(s_{k-1}\right) = \min_{x_1,\ldots,x_T} \sum_{t=k}^{T} g_t\left(s_{t-1}, x_t\right),$$  (4.12)

$$s_t = \phi_t\left(s_{t-1}, x_t\right), t = k, \ldots, T-1.$$

Note that $z = z_1(s_0)$, so by solving (4.12) for $k = 1$ we solve the original problem (4.11) above.

It is not difficult to see that $z_1(s_0)$ can be obtained recursively. A simple rewriting yields what is usually referred to as the *principle of optimality*, that is, in each state a necessary condition for optimality is that the remaining decision are optimal with respect to this state:

$$z_k\left(s_{k-1}\right) = \min_{x_k} \left\{g_k\left(s_{k-1}, x_k\right) + z_{k+1}(s_k)\right\}.$$  (4.13)

This recursion shows that the decisions can be decoupled according to the stage and optimizing at any stage is a single variable problem. We can then sequentially optimize at each stage using backwards recursion, that is, starting with $z_T(s_{T-1})$ we solve $T$ one variable problems in order to eventually obtain $z_1(s_0)$, which solves the original.

$$z_T(s_{T-1}) = \min_{x_T} \left\{g_T\left(s_{T-1}, x_T\right)\right\}$$

$$z_{T-1}(s_{T-2}) = \min_{x_{T-1}} \left\{g_{T-1}\left(s_{T-2}, x_{T-1}\right) + z_T(s_{T-1})\right\}$$

$$\vdots \quad \vdots$$

$$z_1(s_0) = \min_{x_1} \left\{g_1\left(s_0, x_1\right) + z_2(s_1)\right\}.$$

The decoupling of the decision variables and the decomposition of a $T$ variable optimization problem into $T$ single-variable optimization problems is very useful, but it requires that the problem can be reformulated on the form (4.11) above.

The NP method can be thought of in terms of dynamic programming by identifying each stage with a level in the partitioning tree, that is,

$$s_k = \left(x_1^0, \ldots, x_k^0\right),$$  (4.14)

where at level $d$ in the partitioning tree, the selected subregion is defined by

$$\sigma = \left\{x \in X : x_i = x_i^0, i = 1, \ldots, d\right\}.$$

Thus, similar to dynamic programming the NP method can be viewed as fixing one variable at a time. It does however not require that the problem can be reformulated as (4.11). Furthermore, rather than optimally solving the (one-variable) problem at each stage, the NP method solves it heuristically by generating high-quality feasible sample solutions. In other words, instead of equation (4.13), in each step of the NP method we solve a problem of the form

$$\min_{x_k} g\left(s_{k-1}, x_k, x_{k+1}, ..., x_T\right),$$ (4.15)

where the state $s_{k-1}$ is fixed by the partitioning , each possible value of $x_k$ is corresponds to one of the subregion, and the values of the remaining variables $x_{k+1}, ..., x_T$ are randomly generated.

It is also interesting to consider how the NP method can be applied to problems that can be formulated as (4.11) above. In this case fixing variables by the partitioning also fixes the contribution of these variables to the objective function, and since every value of $x_k$ is considered the objective function contribution $g_k\left(s_{k-1}, x_k\right)$ can be calculated. The contribution from the remaining variables $x_{k+1}, ..., x_T$ is unknown but can be estimated based on sample values calculated by randomly assigning values to these remaining variables. Thus, at each step we solve a problem that is closely related to (4.13), namely,

$$z_k\left(s_{k-1}\right) = \max_{x_k}\left\{g_k\left(s_{k-1}, x_k\right) + \tilde{z}_{k+1}(s_k)\right\}.$$ (4.16)

Here the exact solution $z_{k+1}(s_k)$ is replaced by a heuristic solution $\tilde{z}_{k+1}(s_k)$ that is obtained by randomly generating solutions from the region

$$\sigma_j(k-1) = \left\{x \in X : x_i = x_i^0, i = 1, ..., k-1, x_k = x_k^{[j]}\right\},$$

where $x_k^{[j]}$ is the $j$th value of variable $x_k$ and defines the $j$th subregion of region $\sigma(k-1)$.

We conclude from this section that the NP method does have certain similarities to well known exact optimization methods. However, the flexibility of the NP method in incorporating domain knowledge and fast heuristics to generate solution, as well as its use of sampling to deal with difficult constraints, make it better suited for most large-scale complex problems.

In the next three sections we will see how various mathematical programming methods can also be incorporated into all phases of the NP method to improve its efficiency. Specifically, we will show how to use mathematical programming for intelligent partitioning, faster generation of high-quality feasible solutions, and to improve the promising index.

## 4.3 Intelligent Partitioning

In Chapter 2 we discussed how partitioning places a structure on the search space and is hence very important for the efficiency of the search. Intelligent partitioning uses our understanding of the problem to impose a structure that tends to cluster together good solutions. Unfortunately, such intelligent partitioning is certainly not apparent or trivially obtained in most problems that arise in complex applications.

Solving a relaxation may result in sufficient information to construct an intelligent partitioning. Say for example that we are solving a binary integer program (BIP), defined by

$$z_{BIP} = \min_{x} cx \tag{4.17}$$
$$Ax \le b$$
$$x \in \{0, 1\},$$

that is, $X = \{x \in \{0, 1\} : Ax \le b\}$. The LP relaxation of BIP is

$$z_{LP} = \min_{x} cx$$
$$Ax \le b$$
$$x \in [0, 1].$$

This is an easy LP, which can be solved to obtain some optimal solution $x_1^{LP}, x_2^{LP}, ..., x_n^{LP}$. In general $x_i^{LP} \notin \{0, 1\}$ but the value can be taken as an indication of its importance. For example, if $x_i = 0.95$ then it is intuitive that most of the good feasible solutions correspond to $x_i = 1$ and most of the poor feasible solutions correspond to $x_i = 0$. On the other hand, if $x_i = 0.5$ no such inference can be made. One possible intelligent partitioning for the BIP is therefore to order the variables according to the absolute deviation from one half, that is,

$$\left| \frac{1}{2} - x_{[1]} \right| \ge \left| \frac{1}{2} - x_{[2]} \right| ... \ge \left| \frac{1}{2} - x_{[n]} \right| \tag{4.18}$$

and start by partitioning $\sigma(0) = X$ into two subregions

$$\sigma_1(0) = \left\{ x \in \{0, 1\} : x_{[1]} = 0, Ax \le b \right\},$$
$$\sigma_2(0) = \left\{ x \in \{0, 1\} : x_{[1]} = 1, Ax \le b \right\}.$$

We then continue to partition by fixing the remaining variables in the order (4.18) obtained by solving the LP relaxation. It is important to note that such intelligent partitioning is only a heuristic. Is is possible that even though $x_i = 0.95$ in the relaxed solution that $x_i = 0$ in the optimal solution. However, our empirical experience indicates that using such intuitive heuristics for intelligent partitioning is very effective in practice.

Similar to the LP relaxation for the BIP, solving any relaxation will reveal some information about what values are desirable for each variable. This can be utilized for developing an intelligent partitioning but the exact approach will in general depend on the specifics of the application.

We now illustrate the above ideas through a difficult to solve application example, namely the resource-constrained project scheduling problem introduced in Chapter 1. Recall that a project consists of a set of tasks to be performed and a given precedence requirements between some of the tasks.

The project scheduling problem without resource constraints involves finding the starting time of each task so that the overall completion time of the project is minimized. It is well-known that this problem can be solved efficiently using what is called the critical path method that uses forward recursion to find the earliest possible completion time for each task (Pinedo, 2000). The completion time of the last task defines the makespan or the completion time of the entire project.

Now assume that one or more resource is required to complete each task. The resources are limited so if a set of tasks requires more than the available resources they cannot be performed concurrently. The problem now becomes NP-hard and cannot be solved efficiently to optimality using any traditional methods. Using the notation from Chapter 1, the decision variables are the starting times for each task,

$$x_i = \text{Starting time of task } i \in V, \tag{4.19}$$

where $V$ is the set of tasks. We also define the set of tasks processed at time $t$ as

$$V(t) = \{i : x_i \leq t \leq x_i + p_i\},$$

where $p_i$ is the processing time of task $i \in V$. As previously shown in Chapter 1, the resource-constrained project scheduling problem may be formulated mathematically as follows:

$$\min \max_{i \in V} x_i + p_i \tag{4.20}$$

$$x_i + p_i \leq x_j, \ \forall (i,j) \in E \tag{4.21}$$

$$\sum_{i \in V(t)} r_{ik} \leq R_k, \ \forall k \in R, t \in \mathbf{Z}_+^1 \tag{4.22}$$

$$x_i \in \mathbf{Z}_+^1$$

Here $E$ is the set of all precedence constraints, $R$ is the set of resources, $R_k$ is the available resources of type $k \in R$, and $r_{ik}$ is the amount of resources of type $k$ required by task $i \in V$.

It is well known that as noted above the precedence constraints (4.21) are easy, whereas the resource constraints (4.22) are hard. By this we mean that if the constraints (4.22) are dropped then the problem becomes easy to solve using the critical path method. This would hence be an easy to solve relaxation, but unfortunately it is not very useful since it is unlikely to result in useful bounds.

Instead of dropping the difficult resource constraints (4.22), an alternative is to incorporate mathematical programming into the NP framework by considering the continuous relaxation:

$$\min \max_{i \in V} x_i + p_i \tag{4.23}$$

$$x_i + p_i \leq x_j, \ \forall (i,j) \in E \tag{4.24}$$

$$\sum_{i \in V(t)} r_{ik} \leq R_k, \ \forall k \in R, t \in \mathbf{Z}_+^1 \tag{4.25}$$

$$x_i \in \mathbf{R}_+^1$$

The same structure can be used to partition intelligently. Instead of partitioning directly using the decision variables (4.19), we note that it is sufficient to partition to resolve the resource conflicts. Once those are resolved then the problem is solved. This approach is applicable to any problem that can be decomposed in a similar manner.

## 4.4 Generating Feasible Solutions

At first glance it may seem that mathematical programming methods would not be very useful for generating feasible solutions within the NP method. The focus of such methods for discrete problems is the generation of lower bounds that rarely correspond to feasible solutions. However, it turns out that mathematical programming methods can indeed be very useful for generating feasible solutions and incorporating them for this purpose can significantly improve the efficiency of the NP method.

In this chapter we present two distinct ways in which mathematical programming can be used to generate feasible solutions. First, similar to the intelligent partitioning discussed above, the solution to a relaxation of the original problem (4.1) can be used to bias the sampling. The basic idea is for solutions that are similar to the optimal solutions for the relaxed problem to be sampled with higher probability. To illustrate, we consider again the generic BIP discussed above, namely,

$$z_{BIP} = \min_x cx$$

$$Ax \leq b$$

$$x \in \{0, 1\},$$

The LP relaxation can be solved to obtain some optimal solution $x_1^{LP}$, $x_2^{LP}$,..., $x_n^{LP}$ and we can then bias the sampling distribution according to these values. For example, for any $x_i$ we can take the sampling distribution to be

$$P[x_i = 1] = x_i^{LP}, \tag{4.26}$$

$$P[x_i = 0] = 1 - x_i^{LP}. \tag{4.27}$$

Thus, if a particular variable $x_i$ is close to one in the LP relaxation solution then it is one with high probability in the sample solution, and vice versa.

The second approach to incorporating mathematical programming into the generation of feasible solutions applies when the problem (4.1) can be decomposed into two parts, one that is easy from a mathematical programming perspective and one that is hard. For such problem it is impossible in practice to use mathematical programming to solve the entire problem, but when solving the problem using the NP method we can take advantage of the fact that mathematical programming can effectively solve a partial problem. Specifically, we can use sampling to generate partial solutions that fix the hard part of the problem and then complete the solution by solving a mathematical program. Since the mathematical programming output is optimal given the partial sample solution, this process can be expected to result in higher quality feasible solutions than if the entire solution was obtained using sampling. On the other hand, the process of generating a sample solution is still fast since the difficult part of the problem is handled using sampling. This first part can incorporate any biased sampling approach or heuristics, and the combined procedure for generating feasible solutions is therefore a prime example of how mathematical programming and heuristics search compliment each other when both incorporated into the NP framework.

We now return to the resource-constrained project scheduling problem used in the previous section. We know that the precedence constraints (1.4) are easy, whereas the resource constraints (1.5) are hard. By this we mean that if the constraints (1.5) are dropped then the problem becomes easy to solve. As noted before, such problems, where complicating constraints transform the problem from easy to very hard, are common in large-scale optimization.

The flexibility of the NP method allows us to address such problems effectively by taking advantage of special structure when generating feasible solutions. It is important to note that it is very easy to use sampling to generate feasible solutions that satisfy very complicated constraints, which are very difficult to handle using traditional methods such as mathematical programming. Therefore, when faced with a problem with complicating constraints we want to use random sampling to generate partial feasible solutions that resolve the difficult part of the problem and then completed the solution using the appropriate efficient optimization method.

For example, when generating a feasible solution for the resource-constrained project scheduling problem, the resource allocation should be generated using random sampling and the solution can then be completed by applying the critical path method to determine the starting times for each task. This requires reformulating the problem so that the resource and precedence constraints can be separated, but such a reformulation is rather easily achieved by noting that the resource constraints can be resolved by determining a sequence between the tasks that require the same resource(s) at the the same time. Once this sequence is determined then the sequence can be added as easy to solve precedence constraints and the remaining solution generated using the critical path method. Feasible solutions can therefore be generated in the NP method by first randomly sampling a sequence to resolve resource conflicts and then applying the critical path method. Both procedures are very fast so complete sample solutions can be generated rapidly.

We also note that constraints that are difficult for optimization methods such as mathematical programming are sometime very easily addressed in practice by incorporating domain knowledge. For example, a domain expert may easily be able to specify priorities among tasks requiring the same resource(s) in the resource-constrained project scheduling problem. The domain expert can therefore specify some priority rules to convert a very complex problem into an easy one. The NP method can effectively incorporate such domain knowledge into the optimization framework by using the priority rules when generating feasible solutions. This is particularly effective because the domain expert would not need to specify priority rules to resolve all resource conflicts. Rather, any available priority rule or other domain knowledge can be incorporated to guide the sampling.

## 4.5 Promising Index

We recall from Chapter 2 that while the basic method for defining the promising index of a valid region $\sigma \in \Sigma$ is based on the set of feasible solutions $D_\sigma$ that are generated from this region it is possible to incorporate other information about this region. For example, assume that we are solving a general IP and we have a lower bound $\underline{z}(\sigma)$ on the objective function for $\sigma \in \Sigma$. Then this *local lower bound* can be combined with the upper bound $\min_{x \in D_\sigma} cx$ into a single promising index

$$I(\sigma) = \alpha_1 \cdot \underline{z}(\sigma) + \alpha_2 \cdot \min_{x \in D_\sigma} cx, \tag{4.28}$$

where $\alpha_1, \alpha_2 \in \mathbf{R}$ are the weights given to the lower bound and upper bound, respectively. This lower bound can be obtained using any of the techniques discussed in this chapter, such as by solving a LP relaxation, solving the Lagrangian dual, or through an application-specific COP relaxation. Since the promising index now contains more information it may be expected that the correct move is selected more frequently, hence improving the overall efficiency of the NP method.

## 4.6 Non-linear Programming

In this chapter we have studied optimization programs on the form usually assumed by integer programming, namely,

$$z_{IP} = \min_{x \in X} cx,$$

where $X = \{x \in Z_+^n : Ax \leq b\}$. The NP method, however, does not require linearity in either the objective function or the constraints and the ideas presented in this chapter can all be readily extended to *non-linear programming* problems (NLP). In this section we illustrate how the NP method deals with non-linearity in both the objective function and the constraints.

Suppose that we have a general non-linear optimization problem with integer decision variables, that is,

$$\min_{x \in X} f(x),$$ (4.29)

where $X = \{x \in Z_+^n : g(x) \leq b\}$. This is in general a very hard problem. Standard integer programming methods are unlikely to be effective or even applicable due to the non-linearity. Standard non-linear programming methods require a differential objective function. Therefore, the discrete nature of the feasible region makes such methods not directly applicable. On the other hand, since all that is required to apply the NP method is the ability to partition and randomly generate feasible solutions, it can be applied to (4.29) just as it can for (4.1).

While the pure NP method is always applicable for (4.29), its efficiency may often be greatly improved by exploiting special structure. As an extreme case, say for example that the objective function is of the special form

$$f(x) = cx_1 + f'(x_2),$$ (4.30)

where $x = (x_1, x_2)$ and the feasible region is part linear, that is,

$$X = \left\{x_1 \in \mathbf{Z}_+^{n_1}, x_2 \in \mathbf{Z}_+^{n_2} : Ax_1 \leq b_1, g'(x_2) \leq b_2\right\}.$$

In other words, the objective function and constraints are only non-linear through some of the decision variables. In this case the problem becomes considerably easier once the variables $x_2$ have been fixed to some value $x_2^0$, namely it would reduce to the integer program

$$z = \min cx_1 + c_0$$
$$Ax_1 \leq b_1$$
$$x_1 \in \mathbf{Z}_+^{n_1}$$

where $c_0 = f'\left(x_2^0\right)$ is a constant. In many cases it may be possible to solve this reduced problem efficiently using standard integer programming techniques. The NP method can take advantage of this by partitioning on the difficult variables $x_2$, and quickly generating high-quality feasible solutions in two phases: use random sampling to determine the values of the difficult variables $x_2$ and then completing the solution by solving for $x_1$ using standard integer programming methods.

The fact that the NP method does not require linearity in either the objective function or the constraints is very significant. While most, if not all, combinatorial optimization problems can be formulated as MIPs, such linear formulations typically require a large number of variables and/or constraints. In formulating many real problems there is thus often a trade-off between a large formulation where traditional MIP solution methods are applicable and smaller formulations where such methods are not applicable due to non-linearities. When using the NP method it is possible to simultaneously take advantage of the smaller non-linear formulations and MIP methods by incorporating the MIP solution methods to address the linear part of the problem and using sampling to address the non-linear part of the problem. In the application part of the book we will see several occasions where it is beneficial to reformulate linear MIPs as non-linear programs when applying the NP method.

We have seen in this chapter that the NP method has certain connections to standard mathematical programming techniques such as branch and bound and dynamic programming. However, the NP method is primarily useful for problems that are either too large or too complex for mathematical programming to be effective. But even for such problems mathematical programming methods can often be used to solve either a relaxed problem or a subproblem of the original and these solutions can be effectively incorporated into the NP framework in numerous ways. For example, we have seen that this can be done by letting the mathematical programming solution define an intelligent partitioning . It can also be used for generating better feasible solutions, either by biasing the sampling distribution or by using a hybrid technique where the difficult decisions are first determined using sampling and the remaining solution is generating by solving a mathematical program. Finally, the mathematical programming solution can also be incorporated into the promising index.

# 5

# Hybrid Nested Partitions Algorithm

The inherent flexibility of the NP method allows us to incorporate any other heuristic for generating good feasible solutions, resulting in what we call *hybrid Nested Partitions* algorithms. In this chapter we will show how to do this for several of the most popular and effective heuristic search methods, namely genetic algorithms, tabu search, and ant colony optimization, as well as for greedy heuristics.

The basic idea behind effective NP hybrids is to note that in order for the NP method to make a correct move, that is, move either to a subregion containing the global optimum or backtrack, high-quality feasible solutions must be generated from each region being considered. Furthermore, while the NP method very effectively guides the search effort globally, many other heuristics have been shown to be effective in generating an improving sequence of feasible solutions. By incorporating any such algorithm to generate feasible solutions the effectiveness and efficiency of the NP method could be improved. Furthermore, our empirical experience clearly shows that the resulting hybrids are more effective than either the pure NP or the other pure heuristic on their own.

## 5.1 Greedy Heuristics in the NP Framework

The simplest heuristics in terms of implementation are greedy heuristics. We distinguish between two type of greedy heuristics: construction heuristics and improvement heuristics. Construction heuristics build up a single feasible solution by determining the values of the decision variables one-by-one. Improvement heuristics, on the other hand, start with a feasible solution and then attempt to improve it by making some relatively small modification to the solution. An improvement heuristic thus generates a sequence of monotonously improving solutions and terminates when no further improvements are possible, that is, a local optimum has been reached.

In this section we present two ways in which greedy heuristics can be incorporated into the NP framework to improve its efficiency. First, we discuss the use of greedy heuristics for generating feasible solutions, and second, we demonstrate the use of greedy construction heuristics for defining intelligent partitioning.

### 5.1.1 Generating Good Feasible Solutions

As before let $x_i$, $i = 1, 2, ..., n$ denote the decision variables. A greedy construction heuristic fixes the values of the variables one at a time, $x_i = x_i^0$ by using a single variable objective function $g_i(x_i)$, which is believed to correlate in some way with the original objective function defined for $x = (x_1, ..., x_n)$. We can therefore think of a construction heuristic as solving $n$ single variable problems

$$\min_{x_i} g_i(x_i). \tag{5.1}$$

The solution of these $n$ problems will of course not correspond to the optimal solution except for trivial problems.

There are two natural ways of taking advantage of any such construction heuristic within the NP framework: to bias the sampling distribution and to define an intelligent partitioning. The latter issue of intelligent partitioning will be discussed in Section 5.1.2, but first we consider how to use construction heuristics to bias a sampling distribution for randomly generating high-quality solutions. Assume for simplicity, but without loss of generality, that the performance function is non-negative, that is, $g_i(x) \geq 0$, $\forall x$. Then we can define a probability distribution

$$P\left[x_i = x_i^0\right] = \frac{g_i\left(x_i^{max}\right) - g_i\left(x_i^0\right)}{g_i\left(x_i^{max}\right)} \tag{5.2}$$

where

$$x_i^{max} = \mathrm{argmax}_{x_i} g_i(x_i).$$

According to this distribution, good values of each variable will be sampled with higher probability than values that are believed to be poorer. Thus, while any solution still has a change of being selection, the probability is biased so that the overall quality of the sample solutions used by the NP algorithm can be expected to improve. This will in turn make it more likely that the correct solution is selected in each step of the algorithm.

We now turn our attention to heuristics that we call greedy search heuristics or improvement heuristics, that transform a feasible solution $x^0$ into another better solution $\tilde{x}$ through a series of improvement moves from one solution to another similar solution. Specifically, for any point $x \in X$ the greedy search defines a neighborhood $\mathcal{N}(x) \subset X$ of $x$, which is comprised of solutions that are similar to $x$ in some sense. In each iteration of the greedy search, a solution is selected from the neighborhood so that this new solution is better than the previous solution.

Specifically, assume $x^k$ is the current solution (in the $k$th iteration of the search. The greedy search then selects $x^{k+1} \in \mathcal{N}(x^k)$ such that $f\left(x^{k+1}\right) \leq f\left(x^k\right)$, that is, the new solution is an improvement. If no such solution exists the greedy search terminates and the current solution is a local optimum given the neighborhood structure.

A greedy local search is typically very fast but obviously suffers from the limitation that it terminates at the first local optimum it encounters. However, when incorporated into the NP framework to generate good feasible solutions the benefits of fast greedy search and global convergence of the NP framework are combined into what usually turns out to be a more efficient NP method. This is illustrated in the following example.

**Product Design Problem**

We will demonstrate the use of greedy local search in the NP method for the *product design problem*. Such optimization problems occur when designing new products to satisfy the preferences of expected customers. These problems may be divided into single product design problems where the objective is to design the attributes of a single product (Kohli and Krishnamurti 1989, Balakrishnan and Jacob 1996), and product line design where multiple products are offered simultaneously (Green and Krieger 1985, Kohli and Sukumar 1990, Dobson and Kalish 1993, Nair et al. 1995). Here we focus on the single product design problem that involves determining the levels of the attributes of a new or redesigned product in such a way that it maximizes a given objective function. We assume that the preferences of individual customers, or market segment, have been elicited for each level of every attribute (Zufryden 1977, Green et al. 1981, Green and Srinivasan 1990). Furthermore, we assume that all product designs found by combining different levels of each attribute are technologically and economically feasible, and that a customer will choose the offered product if its utility is higher than that of a competing status quo product, which may be different for each customer. This problem is usually referred to as the *share-of-choices problem*.

The share-of-choices problem is very difficult to solve, especially as the product complexity increases and more attributes are introduced. In fact, it belongs to the class of NP-hard problems (Kohli and Krishnamurti 1989), so the exact mathematical programming methods such as branch-and-bound (see Section 4.1) are not likely to be successful for realistically sized problems. On the other hand, the NP method is clearly applicable.

We assume that a product has $K$ attributes, each of which can be set to one of $L_k$ levels, $k = 1, ..., K$. To formulate the share-of-choices problem we define the decision variables as $x = \{x_{kl}\}$, where for each $l \in \{1, 2, \ldots, L_k\}$, $k \in \{1, 2, \ldots, K\}$, $x_{kl} = 1$, if attribute $k$ is set to level $l$. Otherwise $x_{kl} = 0$. The objective is to maximize the market share and a customer $i$ will purchase the offered product if and only if this customers total utility, $TU_i(x)$, for this product is greater than the total utility $q_i$ of this customers status quo product, $1 \leq i \leq M$. The total utility the $i$th customer obtains from the product is

$$TU_i(x) = \sum_{k=1}^{K} \sum_{l=1}^{L_k} u_{ikl} \cdot x_{kl},$$

where $u_{ikl}$ is the utility the $i$th customer derives if attribute $k$ is set to level $l$, $i = 1, ..., N$. These utilities can be represented as parts-worth matrices for each attribute $k$:

$$U(k) = \begin{bmatrix} u_{1k1} & \cdots & u_{1kL_1} \\ \vdots & \ddots & \vdots \\ u_{Nk1} & \cdots & u_{NkL_N} \end{bmatrix},$$

where $N$ is the total number of customers. Each customer purchases the offered product if and only if $TU_i(x) > q_i$, so the performance function for the $i$th customer can be written as

$$f_i(x) = \frac{\max\{0, TU_i(x) - q_i\}}{TU_i(x) - q_i}, \tag{5.3}$$

when $TU_i(x) \neq q_i$ and $f_i(x) = 0$ when $TU_i(x) = q_i$. This performance function assigns one to all the customers that purchase the offered product and zero to all other customers. Now the share-of-choices problem may be stated as the following mathematical programming problem,

$$\max_{x} f(x) = \sum_{i=1}^{M} f_i(x), \tag{5.4}$$

subject to

$$\sum_{l=1}^{L_k} x_{kl} = 1, \quad k = 1, 2, ..., K \tag{5.5}$$

$$x_{kl} \in \{0, 1\}, \quad \forall k, l. \tag{5.6}$$

In the above notation, a product profile $x$ is a string of length $\sum_{k=1}^{K} L_k$, that consists only of zeros and ones. However, since there are only $K$ ones, indicating which level is chosen for each attribute, and the reminder of the string is zeros, we can represent a product profile more compactly as a string of length $K$, denoted $x = [l_1 \ l_2 \ \cdots \ l_K]$. Each element of this vector represents one attribute, and the value of the element indicates which level is chosen. For example, attribute $k$ is set to level $l_k$, where $1 \leq k \leq K$. This simplifies the notation.

## NP for Product Design

To illustrate the NP method for product design, consider a small problem with $K = 3$ attributes, $L_1 = 3$, and $L_2 = L_3 = 2$. For some $N = 3$ individuals, let the part-worths data matrices be as follows:

$$U(1) = \begin{bmatrix} 1 & 1 & 3 \\ 1 & 4 & 1 \\ 2 & 2 & 0 \end{bmatrix}, \qquad U(2) = \begin{bmatrix} 3 & 0 \\ 1 & 2 \\ 2 & 2 \end{bmatrix}, \qquad U(3) = \begin{bmatrix} 1 & 1 \\ 1 & 4 \\ 3 & 1 \end{bmatrix}.$$

Suppose the status-quo is represented by the values of level 1 for each attribute. Then subtracting the first column from each column of gives the relative parts-worth data matrices:

$$\tilde{U}(1) = \begin{bmatrix} 0 & 0 & 2 \\ 0 & 3 & 0 \\ 0 & 0 & -2 \end{bmatrix}, \qquad \tilde{U}(2) = \begin{bmatrix} 0 & -3 \\ 0 & 1 \\ 0 & 0 \end{bmatrix}, \qquad \tilde{U}(3) = \begin{bmatrix} 0 & 0 \\ 0 & 3 \\ 0 & -2 \end{bmatrix}.$$

This allows us to calculate the performance of each design, that is, the expected number of customers selecting the product, according to equation (5.3) above. In this chapter we refer to this product design problem as the 3-Attribute example.

This example has 22 valid regions. The region containing all product profiles $X \in \Sigma$ has three subregions $\sigma_1$, $\sigma_2$, and $\sigma_3$; and each of these subregions has two subregions of its own. The maximum depth is $d^* = 3$ and there are 12 regions of maximum depth, that is $|\Sigma_0| = 12$. Notice that these singleton regions define a complete product profile, whereas every region $\eta \in \Sigma$ such that $0 < d(\eta) < d^*$ defines a partial product profile. Also note that the maximum depth will always be equal to the number of attributes. The best feasible solution value is initialized as $f^0 = 0$.

The pure NP algorithm starts from the initial most promising region, $X$, and must determine which of its three subregions:

$$\sigma_1 = \{[l_1 \; l_2 \; l_3] : l_1 = 1, 1 \leq l_2 \leq 2, 1 \leq l_3 \leq 2\},$$
$$\sigma_2 = \{[l_1 \; l_2 \; l_3] : l_1 = 2, 1 \leq l_2 \leq 2, 1 \leq l_3 \leq 2\},$$
$$\sigma_3 = \{[l_1 \; l_2 \; l_3] : l_1 = 3, 1 \leq l_2 \leq 2, 1 \leq l_3 \leq 2\},$$

will become the most promising region in the next iteration. To determine this region, each of these subregions is sampled by selecting the values of $l_2$ and $l_3$. This involves two step sampling. In the first step we generate a uniform random variable $u \in (0,1)$. If $u < \frac{1}{2}$ then we set $l_2 = 1$, and if $\frac{1}{2} \leq u < 1$ then we set $l_2 = 2$. In the second step we generate another uniform random variable $v \in (0,1)$. If $v < \frac{1}{2}$ then we set $l_3 = 1$, if $\frac{1}{2} \leq v < 1$ then we set $l_3 = 2$. For example, when we sample in $\sigma_1$, if $u < \frac{1}{2}$ in the first step and $\frac{1}{2} \leq v < 1$ in the second step, then the sample product profile generated is $x^1 = \begin{bmatrix} l_1^1 & l_2^1 & l_3^1 \end{bmatrix} = [1 \; 1 \; 2]$. This procedure is identical for each of the three regions. Suppose the other two samples from $\sigma_2$ and $\sigma_3$ are $x^2 = \begin{bmatrix} l_1^2 & l_2^2 & l_3^2 \end{bmatrix} = [2\ 1\ 1]$ and $x^3 = \begin{bmatrix} l_1^3 & l_2^3 & l_3^3 \end{bmatrix} = [3\ 2\ 2]$, respectively.

After we obtain one sample product profile from each region currently under consideration, the next step is to estimate the promising index of each region using the sample product profiles. Usually the number of samples in each region should be larger than one. However, for ease of exposition, in this example we use single sample in each region and its objective value as estimated promising index: $\hat{I}(\sigma_1) = f(x^1) = f([1\ 1\ 2]) = 1$, $\hat{I}(\sigma_2) = f(x^2) = f([2\ 1\ 1]) = 1$, $\hat{I}(\sigma_3) = f(x^3) = f([3\ 2\ 2]) = 1$. By breaking the tie arbitrarily the algorithm moves to region $\sigma_1 = \{[l_1\ l_2\ l_3] : l_1 = 1, 1 \le l_2 \le 2, 1 \le l_3 \le 2\}$ and we update the best solution value to $f^0 = 1$.

As $\sigma_1$ is now the most promising region, the algorithm then samples the subregions

$$\sigma_{11} = \{[l_1\ l_2\ l_3] : l_1 = 1, l_2 = 1, 1 \le l_3 \le 2\}$$
$$\sigma_{12} = \{[l_1\ l_2\ l_3] : l_1 = 1, l_2 = 2, 1 \le l_3 \le 2\}$$

as well as the complimentary region $X \setminus \sigma_1$. Suppose the complimentary region $X \setminus \sigma_1$ has the best estimated promising index, the algorithm backtracks to a larger region. Recall that the algorithm backtracks to a region containing the best product profile found in this iteration and with depth of $\lfloor \frac{d^*}{2} \rfloor + 1$, where $\lfloor x \rfloor$ denotes the largest integer smaller than or equal to $x$, for any $x \in \mathbf{R}$. Therefore, the most promising region will be set to $\sigma_{31}$, and the best solution $f^0 = 2$ is updated. It can be shown by enumeration that 2 customers is the optimal solution for this data set. The algorithm will continue in this manner until some predefined stopping criteria is met.

## Comparison of Hybrid and Non-Hybrid Methods

Greedy search heuristic have previously been proposed for the share-of choices problem (Kohli and Krishnamurti 1989). In this section we define a greedy search (GS) heuristic as follows: For each attribute $k$, the overall relative utilities for all the customers is calculated, that is

$$u_{kl} = \sum_{i=1}^{M} u_{ikl} - u_{ik\tilde{l}_k},$$

where $\tilde{l}_k$ is the level of the $k$th attribute for the status-quo product. The GS heuristic then selects the level $l_k^*$ for the $k$th attribute that maximizes the relative utilities. Thus the level of each attribute is determined independently of all other attributes.

We now compare the effectiveness of the pure NP method with uniform random sampling used to generate feasible solution and the pure greedy search for the product design problem to the hybrid NP/GS algorithm of using greedy

**Table 5.1.** Comparison of pure NP, greedy search, and hybrid NP/GS.

| Problem | | GS | NP-Pure | NP/GS |
|---|---|---|---|---|
| $K$ | $L$ | | | |
| 5 | 5 | 219 | 238 | 238 |
| 6 | 6 | 237 | 239 | 239 |
| 7 | 7 | 225 | 239 | 241 |
| 8 | 8 | 228 | 236 | 239 |
| 9 | 9 | 223 | 240 | 244 |
| 10 | 10 | 241 | 241 | 248 |
| 20 | 20 | 237 | 248 | 256 |

search for generating higher quality solutions from each region in the NP method. The results are reported in Table 5.1.

From these results we see that the pure NP method consistently finds better solutions than the greedy search algorithm. Of course this should be expected since the greedy search is also much faster and simply terminates at the first local optimum. The more interesting comparison is between the pure NP method and the hybrid NP/GS method that incorporates greedy search into the solution generation. For the two smallest problems the solution quality is the same but for each of the other problems the hybrid algorithm finds better solutions. Furthermore, the gap between the solution quality increases with the size of the problem. These results therefore indicate that the hybrid NP algorithm performs best for large-scale problems. As we will see throughout this chapter and the remainder of the book, this is a consistent observation for different applications and different NP hybrids. When compared to other methods, hybrid NP algorithms perform relatively best when applied to solve large-scale complex problems.

In the application part of this book we will see numerous other examples of how to effectively incorporate greedy local search into the NP method to improve its efficiency, but next we consider how to incorporate greedy heuristics into the partitioning.

### 5.1.2 Intelligent Partitioning

Many construction heuristics use an ordering of the importance of the variables to construct a good solution. This is for example true of the classic shifted-bottleneck heuristic for job-shop scheduling (Pinedo, 2000). The job-shop scheduling problem involves scheduling a set of jobs on a set of machines. Each job must be processed on a given subset of machines in a given order. There are therefore two sets of constraints: (a) precedence constraints that assure that each job follows the correct route through the job shop, and (b) machine constraints that assure that each machine only processes one job at a time. The shifted-bottleneck heuristic orders the machines according to their importance and first sequences all of the jobs that must be processed on the most important (bottleneck) machine, then sequences all the jobs on the second most important machine, and so forth.

The same principle used by the shifted-bottleneck heuristic can be used to impose a structure on the search space when partitioning in the NP method. When solving the job-shop scheduling problem the variables corresponding to the bottleneck machine should therefore be partitioned on first, and so forth. In general, when partitioning for any problem the most important variables should be used at the top of the partitioning tree. For a given application, any available heuristic that measures the importance of variables can be applied for this purpose and in many applications there are multiple ways of measuring the importance of variables. Finally, we note that a similar principle for branching variable selection is well-known to perform well when applying the branch-and-bound algorithm for obtaining exact solutions (Nemhauser and Wolsey 1988).

We will revisit the issue of using heuristics for intelligent partitioning in Chapter 7, where we show that by using very simple construction heuristics the speed of the NP algorithm can be improved by an order of magnitude.

## 5.2 Random Search in the NP Framework

The greedy heuristics discussed in the previous section are typically very fast but limited in that they only explore a small portion of the feasible region before terminating at a local optimum. The solution quality for problems with complex structure is therefore often not satisfactory. On the other end of the spectrum are the exact mathematical programming methods discussed in Chapter 4. These methods guarantee an optimal solution but the solution time is usually too long in practice. For problems where greedy search heuristics result in unsatisfactory solutions and mathematical programming methods are too time consuming, random search methods and metaheuristics have been found to be very effective in practice (Gendreau and Potvin 2005, Glover and Kochenberger 2003, Lovsz 1996). Such methods usually use one or more local (greedy) improvement moves at their core but use randomization to escape local optima and explore a larger part of the feasible region. The search therefore generates a sequence of feasible solutions that are heuristically believed to be good but are not necessarily monotonously improving as for greedy search.

Similar to the greedy search discussed above, any random search heuristic can be incorporated into the NP method by using it to generate feasible solutions from the regions under consideration. Such a process may be expected to generate higher quality solutions than pure random sampling, which in turn often increases the probability that the correct region is found to be best and the NP method moves in the correct direction.

There are two basic approaches to utilizing random search methods for generating good feasible solutions in the NP method, depending on if the random search is point-to-point or population-based. Point-to-point random search methods start from a single solution $x^i$ and explore what is usually referred to as a neighborhood $\mathcal{N}\left(x^i\right) \subset X$ of this solution, namely some small set of solutions that are similar according to some distance measure. The method then selects a candidate $x^c \in \mathcal{N}\left(x^i\right)$ from the neighborhood and either accepts it (and $x^{i+1} = x^c$) or rejects it (and sets $x^{i+1} = x^i$). The process is then repeated using the new solution $x^{i+1}$. As mentioned in Chapter 2 above, a natural way to take advantage of any point-to-point random search method within the NP method is to start with a random set $D \subset X$ of sample points, and then use each solution $x \in D$ as the starting point $x^0$ of a random search. Thus, the entire randomly generated set is transformed through the local search into a new set of solutions with better performance. This is repeated for each subregion and the complimentary region, and these new sets of solutions can then be used as the basis for selecting a new promising region. It is intuitively appealing that it is more likely that the correct move is made on the basis of the improved solution sets than the original randomly generated solution sets. This is supported by our empirical experience reported in subsequent chapters.

An even more natural match is achieved through NP hybrids that utilize population based random search methods. Such methods start with a set of solutions $D \subset X$ or a population, which just as in the NP method is typically randomly generated. This set is then improved through a series of operations. Similar to above, we can therefore randomly generate an initial population from each region being considered, apply the search method to each population, and then use the final population from each region to choose the next most-promising region. As before, this may be expected to increase the probability of correct selection and this is indeed also supported by our empirical experience.

In the next three subsection we present three examples to illustrate how random search methods can be incorporated into the NP method for very effective hybrid NP algorithms.

### 5.2.1 NP with Genetic Algorithm

We first consider the popular genetic algorithms and how they can be incorporated into the NP framework for a hybrid NP algorithm. To illustrate the effectiveness of this hybrid approach we use it solve a complex combinatorial optimization problem, namely the product design problem introduced in Section 5.1 above.

### Genetic Algorithm

One of the most popular class of random search methods is the genetic algorithm (GA) and other evolutionary search algorithms (Goldberg 1989, Leipins and Hillard 1989, Muhlenbien 1997). The GA is a population based random search method based on the concept of natural selection. It starts from an initial population and then uses a mixture of reproduction, crossover, and mutation to create new, and hopefully better, populations. The GA usually works with an encoded feasible region, where each feasible point is represented as a string, which is commonly referred to as a chromosome. Each chromosome consists of a number of smaller strings called genes. The reproduction involves selecting the best, or fittest, chromosomes from the current population, the crossover involves generating new chromosomes from this selection, and finally mutation is used to increase variety by making small changes to the genes of a given chromosome.

The genetic algorithm can be though of as an improvement heuristic, that is, it starts out with a set of solutions called the initial population and improves these solutions iteratively, with the goal of improving the population in each iteration. As for other such population-based improvement heuristics, GA can be incorporated into the NP algorithm as follows: Starting with the sample solution sets from each region as initial population, the GA search may be used to improve this population, that is, generate better feasible solutions, and the promising index for each region can then be estimated from the final population of solutions rather than the initial population.

### Hybrid NP/GA Algorithm

More precisely, the following procedure is employed in a *Hybrid NP/GA algorithm*. Suppose a region $\sigma \in \Sigma$ is among those currently being considered. Once an initial population $D^0$ has been obtained from the region then the GA search proceeds as follows: First the $\frac{N}{2}$ best (fittest) solutions are selected for reproduction from the population. These solutions will also survive intact in the next population. Secondly, pairs of solutions are selected randomly from the reproduced solutions. Each variable, that is not fixed in $\sigma$, of the selected pair has a given probability of participating in a crossover, that is, being swapped with the corresponding variable of the other solution. Finally, each solution may be selected with a given probability as a candidate for mutation. If a solution is mutated a variable from that solution is selected at random and assigned a random level. As before, only variables that are not fixed in $\sigma$ may be selected to be mutated.

The hybrid procedure described above can be implemented in the following algorithm.

**Algorithm** *Hybrid NP/GA*

Step 0    **Initialization.** Set $k = 0$ and $\sigma(k) = X$.

Step 1    **Partitioning.** If $d(\sigma(k)) \neq d^*$, that is, $\sigma(k) \notin \Sigma_0$, partition the fittest region, $\sigma(k)$, into $M_{\sigma(k)}$ subregions $\sigma_1(k), ..., \sigma_{M_{\sigma(k)}}(k)$. If $d(\sigma(k)) = d^*$ then let $M_{\sigma(k)} = 1$ and $\sigma_1(k) = \sigma(k)$.

If $d(\sigma(k)) \neq 0$, that is, $\sigma(k) \neq \Theta$, aggregate the complimentary region $\Theta \setminus \sigma(k)$ into one region $\sigma_{M_{\sigma(k)}+1}(k)$.

Step 2 **Initial Population.** If $d(\sigma(k)) \neq d^*$ use a randomized method to obtain an initial population of $N_j$ strings from each of the regions $\sigma_j(k)$, $j = 1, 2, ..., M_{\sigma(k)} + 1$,

$$POP_I^j = \left[ x_I^{j1}, x_I^{j2}, ..., x_I^{jN_j} \right], \quad j = 1, 2, ..., M_{\sigma(k)} + 1. \quad (5.7)$$

If $d(\sigma(k)) = d^*$ then obtain a population $POP_{I_0}^j$ of $N_j - M$ strings as above and let the initial population be $POP_I^j = POP_{I_0}^j \cup POP_0$, where $POP_0$ is the diverse high quality population found in the first iteration (see Step 6). Note that we implicitly assume that $N_j > M$.

Step 3 **GA Search.** Apply the GA to each initial population $POP_I^j$ individually, obtaining a final population for each region $\sigma_j(k)$, $j = 1, 2, ..., M_{\sigma(k)} + 1$,

$$POP_F^j = \left[ x_F^{j1}, x_F^{j2}, ..., x_F^{jN_j} \right], \quad j = 1, 2, ..., M_{\sigma(k)} + 1. \quad (5.8)$$

The number of steps in the GA search can be determined by the progress of the search. As opposed to a pure GA algorithm, there is no need to exhaust the search, but rather the GA search should be terminated when it is no longer making significant improvements. If $k = 0$ then go to Step 6, otherwise go to Step 4.

Step 4 **Overall Fitness.** Estimate the *overall fitness* of the region as the performance of the fittest chromosome in the final population. That is, the overall fitness of each region is estimated by

$$\hat{F}(\sigma_j) = \max_{i \in \{1,2,...,N_j\}} f(x_F^{ji}), \quad j = 1, 2, ..., M_{\sigma(k)} + 1. \quad (5.9)$$

Step 5 **Update Fittest Region.** Calculate the index of the region with the best overall fitness,

$$\hat{j}_k \in \arg \max_{j \in \{1,2,...,M_{\sigma(k)}+1\}} \hat{F}(\sigma_j). \quad (5.10)$$

If more than one region is equally fit, the tie can be broken arbitrarily, except at maximum depth where ties are broken by staying in the current fittest region. If this index corresponds to a region that is a subregion of $\sigma(k)$, then let this be the fittest region in the next iteration. That is $\sigma(k+1) = \sigma_{\hat{j}_k}(k)$ if $\hat{j}_k \leq M_{\sigma(k)}$.

If the index corresponds to the complimentary region, backtrack to a region $\eta \in \Sigma$ that is defined by containing the fittest chromosome in the complimentary region and being of depth $\Delta$ less than the current most promising region. Note that this fittest chromosome

$x_{fit}$ is known as the argument where the minimum (5.9) is realized for $j = M_{\sigma(k)} + 1$. In other words, $\sigma(k+1) = \eta$ where

$$d(\eta) = d(\sigma(k)) - \Delta, \tag{5.11}$$

and

$$x_{fit} \in \eta. \tag{5.12}$$

This uniquely defines the next most promising region.
Let $k = k + 1$. Go back to Step 1.

Step 6    **Initial Diverse Population.** Let $x^j_{\hat{i}_j}$ be the fittest chromosome from the $j$th region

$$\hat{i}_j = \arg \max_{i \in \{1,2,...,N_j\}} x^{ji}_F, \tag{5.13}$$

for $j = 1, 2, ..., M$. Let $POP_0$ be the set of the fittest chromosome from each region

$$POP_0 = \left[ x^{1\hat{i}_1}_F, ..., x^{M\hat{i}_M}_F \right]. \tag{5.14}$$

Go back to Step 4.

In the next section the performance of this hybrid NP/GA algorithm will be compared to the corresponding non-hybrid algorithms.

**Comparison of Hybrid and Non-Hybrid Methods**

We demonstrate the effectiveness of using GA to generate feasible solutions within the NP algorithm through numerical results for the product design problem. These numerical results are for seven different problems. Five small to moderately sized problems and two large problems. Due to the size of most of these problems it is not possible to solve them exactly within reasonable amount of time and the true optimum is therefore unknown. All the problems have $N = 400$ customers and for simplicity we let all the attributes have the same number of levels, that is, $L_k = L$ for all $k = 1, 2, .., K$. The part-worths preferences for each level of each attribute and the status quo prices are generated uniformly for each customer. The details on how the simulated data sets is generated can be found in Nair, Thakur, and Wen (1995). All of the NP algorithms use $N = 20$ sample points from each region and our numerical experience indicates that the performance of the algorithms is fairly insensitive to this number.

For the hybrid NP/GA algorithm a total of ten GA search steps were used in each region. Both the NP/GA and the pure GA had a 20% mutation rate. The performance of each algorithm after a fixed CPU time is given in Table 5.2 for ten replications of each algorithm. These results indicate that it is beneficial to incorporate the GA heuristic into the NP framework. The resulting hybrid algorithm performs no worse than the pure counterpart heuristics in all cases and for large problems there is a substantial improvement.

**Table 5.2.** Performance of Hybrid NP/GA for product design problem.

| Problem | | GA | NP | NP/GA |
|---|---|---|---|---|
| $K$ | $L$ | | | |
| 5 | 5 | 238 | 238 | 238 |
| 6 | 6 | 239 | 239 | 239 |
| 7 | 7 | 241 | 239 | 241 |
| 8 | 8 | 240 | 236 | 240 |
| 9 | 9 | 245 | 240 | 247 |
| 10 | 10 | 250 | 241 | 252 |
| 20 | 20 | 261 | 248 | 266 |

### 5.2.2 NP with Tabu Search

Another widely used random search method is the *tabu search* metaheuristic. In this section we illustrate how to incorporate tabu search into the NP method.

### Tabu Search

Tabu search was introduced by Glover (1989, 1990) to solve combinatorial optimization problems and it has been used effectively for may practical problems. It is a point-to-point method and the main idea is to make certain moves or solutions tabu, that is they cannot be visited as long as they are on the tabu list. The tabu list is dynamic and after each move the latest solution or the move that resulted in this solution is added to the list and the oldest solution or move is removed from the list. Another defining characteristic of tabu search is that the search always selects the best non-tabu solution from the neighborhood of the current solution, even if it is worse than the current solution. This allows the search to escape local optima and the tabu list ensures that the search does not revert back. Tabu search numerous other elements, such as long-term memory that restarts the search, with a news tabu search, at previously found high-quality solutions, and a comprehensive treatment of this methodology can be found in Glover and Laguna (1997).

### Hybrid NP/Tabu Algorithm

Similar to the hybrid NP/GA algorithm above, tabu search can be incorporated into the NP framework by using tabu search to generate feasible sample solutions from each promising region. The results in the following hybrid NP/Tabu algorithm.

**Algorithm** *Hybrid NP/Tabu*

1. **Partitioning.** Partition the most promising region $\sigma(k)$, into $M$ subregions $\sigma_1(k), ..., \sigma_M(k)$, and aggregate the complimentary region $X \setminus \sigma(k)$ into one region $\sigma_{M+1}(k)$.
2. **Random Sampling.** For each region $\sigma_j(k)$, generate $N_j$ feasible sample solutions by repeating the following tabu search procedure $N_j$ times, $j = 1, 2, ..., M + 1$.
   a) Randomly generate a starting point $x_j^0$. Initialize $n = 0$, the current solution $x^n$ and the best solution found so far $x^{best}$ are initialized to the random starting point, that is, $x^{best} = x^n = x_j^0$. The tabu list $L(n)$ is initialized as empty, that is, $L(n) = \emptyset$.
   b) Given a neighborhood $\mathcal{N}(x^n)$ around the current feasible solution, search all feasible neighborhood solutions to find the best non-tabu solution $\tilde{x}^{nontabu}$ and tabu solution $\tilde{x}^{tabu}$ and their respective objective function values $f(\tilde{x}^{nontabu})$ and $f(\tilde{x}^{tabu})$.
   c) We have the following three cases:
      i. If $f(x^{best}) \geq \max\left\{f(\tilde{x}^{nontabu}), f(\tilde{x}^{tabu})\right\}$ let

      $$x^{n+1} = x\tilde{x}^{nontabu}, \qquad (5.15)$$

      and the move $\tilde{x}^{nontabu} \to x^n$ is added to the tabu list

      $$L(n+1) = L(n) \cup \left\{\tilde{x}^{nontabu} \to x^n\right\}. \qquad (5.16)$$

      ii. If $f(\tilde{x}^{nontabu}) \geq \max\left\{f(x^{best}), f(\tilde{x}^{tabu})\right\}$, let

      $$x^{n+1} = \tilde{x}^{nontabu}, \qquad (5.17)$$

      and $x^{best} = x^{n+1}$, $f^{best} = f(x^{n+1})$. The move $\tilde{x}^{nontabu} \to x^n$ is added to the tabu list

      $$L(n+1) = L(n) \cup \left\{\tilde{x}^{nontabu} \to x^n\right\}. \qquad (5.18)$$

      iii. If $f(\tilde{x}^{tabu}) > \max\left\{f(x^{best}), f(\tilde{x}^{nontabu})\right\}$, then let

      $$x^{n+1} = \tilde{x}^{tabu}, \qquad (5.19)$$

      $x^{best} = x^{n+1}$, $f^{best} = f(x^{n+1})$, and the move $\tilde{x}^{tabu} \to x^n$ is added to the tabu list

      $$L(n+1) = L(n) \cup \left\{\tilde{x}^{tabu} \to x^n\right\}. \qquad (5.20)$$

      If the list length exceeds its maximum, remove the oldest move from the tabu list.

   d) If stopping criterion is satisfied, return $x^{best}$ as the next feasible sample solution generated. Otherwise, let $n = n + 1$, return to Step *b*.

After repeating this tabu search process $N_j$ times for each region $\sigma_j(k)$, there are $N_j$ sample solutions from each of the regions $\sigma_j(k)$, $j = 1, 2, ..., M + 1$:

$$x_1^j, x_2^j, ..., x_{N_j}^j, \quad j = 1, 2, ..., M + 1.$$

Calculate the corresponding performance values:

$$f(x_1^j), f(x_2^j), ..., f(x_{N_j}^j), \quad j = 1, 2, ..., M + 1.$$

3. **Calculate promising index.** For each region $\sigma_j$, $j = 1, 2, ..., M + 1$, calculate the promising index as the best performance value within the region:

$$I(\sigma_j) = \min_{i=1,2,...,N_j} f(x_i^j), \quad j = 1, 2, ..., M + 1. \tag{5.21}$$

4. **Move.** Calculate the index of the region with the best performance value.

$$\hat{j}_k \in \arg \min_{j=1,...,M+1} I(\sigma_j), \quad j = 1, 2, ..., M + 1. \tag{5.22}$$

If more than one region is equally promising, the tie can be broken arbitrarily. If this index corresponds to a region that is a subregion of $\sigma(k)$, that is $\hat{j}_k \le M$, then let this be the most promising region in the next iteration

$$\sigma(k + 1) = \sigma_{\hat{j}_k}(k) \tag{5.23}$$

Otherwise, if the index corresponds to the complimentary region, that is $\hat{j}_k = M + 1$, backtrack to the previous most promising region:

$$\sigma(k + 1) = \sigma(k - 1). \tag{5.24}$$

The above hybrid NP/Tabu algorithm will be implemented for a buffer allocation problem in Chapter 12, and the numerical results reported in that chapter will illustrate its effectiveness.

### 5.2.3 NP with Ant Colony Optimization

As a final illustration of incorporating a metaheuristic into the NP framework, we consider ant colony optimization (ACO). Specifically, we show how to develop a hybrid NP/ACO algorithm to solve the traveling salesman problem (TSP) introduced in Chapter 2 above.

### Ant Colony Optimization

Ant colony optimization algorithms imitate the natural behavior of ants in finding the shortest distance between their nests and food sources. Ants exchange information about good routes through a chemical substance called pheromone that accumulates for short routes and evaporate for long routes.

Thus, to imitate this behavior, an algorithm can be defined that identifies what solutions should be visited based on some pheromone values $\tau$, which are updated according to solution quality, and an evaporation rate $\rho < 0$.

The first ACO algorithm is the ant system (AS) of Dorigo (1996) that did not show good results with respect to solving large TSPs. Numerous modifications to the original AS have been introduced to improve the quality of the solution such as having local and global updating of the pheromone matrix as in the ant colony system (ACS) of Dorigo and Gambardella (1997) or allowing certain ants only to update the pheromone matrix. The Max-Min Ant System (MMAS) modifies the AS by keeping the pheromone values within a range $[\tau_{max}, \tau_{min}]$ to ensure that there is an upper bound to the probability that a certain edge is selected (Stützle and Hoos 2000).

### Hybrid NP/ACO Algorithms

Ant colony optimization can be incorporated in the NP framework in a very similar manner to genetic algorithms and tabu search, namely by using ants to generate higher quality feasible solutions from each region.

We illustrate such a hybrid NP/ACO algorithm for solving the TSP. Recall from Chapter 2 that for the TSP there is a set of cities, and a tour must be determined that visits each city exactly once and starts and ends with the same city. The objective is to minimize the length of the tour. Partitioning for the TSP is discussed in detail in Section 2.2 and here we focus only on how to better generate feasible solutions from each region. Specifically, to the following hybrid NP/ACO algorithm, which is based on the work of Al-Shihabi (2004), uses ants to generate high-quality feasible solutions.

### Algorithm *Hybrid NP/ACO*

1. **Partitioning.** Let $L(k)$ denote the list of cities that have already been fixed by the partitioning, that is, $L(k) = \{\tilde{x}_1, ..., \tilde{x}_l\}$, where $\tilde{x}_i$ denotes the city fixed as the $i$th city. Thus, the most promising region is

$$\sigma(k) = \{x \in X | x_i = \tilde{x}_i, i = 1, ..., l\}.$$

Partition $\sigma(k)$ into $M$ subregions $\sigma_1(k), ..., \sigma_M(k)$, where

$$\sigma_j(k) = \left\{ x \in X | x_i = \tilde{x}_i, i = 1, ..., l, x_{l+1} = \tilde{x}_i^{(j)} \right\}.$$

Aggregate the complimentary region $X \setminus \sigma(k)$ into one region $\sigma_{M+1}(k)$.

2. **Random Sampling.** A two part process is used to generate random sample solutions. First, ants are use to find certain high-quality solutions, and then those solutions are perturbed to generate further feasible solutions. First, to generate sample solutions using ants, each ant has a list

$$L^{(j)}(k) = L(k) \cup \{\tilde{x}_i^{(j)}\}$$

of cities already visited, and then when at city $x_i^{(j)}$, the and chooses to visit City $h$ next with probability

$$P_{x_i^{(j)} h} = \frac{\tau_{x_i^{(j)} h}^{\alpha} \cdot \eta_{x_i^{(j)} h}^{\beta}}{\sum_{l \in L^{(j)}(k)} \tau_{x_i^{(j)} h}^{\alpha} \cdot \eta_{x_i^{(j)} h}^{\beta}}. \tag{5.25}$$

The pheromone matrix $\{\tau_{ij}\}$ is updated according to

$$\tau_{ij} = \rho\tau_{ij} + \frac{1}{L_{sa}} \cdot (\text{edge}(i,j) \in) \tag{5.26}$$

After a number of solutions has been generated using this probability, further solutions are generated by perturbing the existing solutions. The end result is $N_j$ randomly generated sample solutions from each of the regions $\sigma_j(k)$, $j = 1, 2, ..., M+1$:

$$x_1^j, x_2^j, ..., x_{N_j}^j, \quad j = 1, 2, ..., M+1,$$

with performance values

$$f(x_1^j), f(x_2^j), ..., f(x_{N_j}^j), \quad j = 1, 2, ..., M+1.$$

3. **Calculate promising index.** For each region $\sigma_j$, $j = 1, 2, ..., M+1$, calculate the promising index as the best performance value within the region:

$$I(\sigma_j) = \min_{i=1,2,...,N_j} f(x_i^j), \quad j = 1, 2, ..., M+1. \tag{5.27}$$

4. **Move.** Calculate the index of the region with the best performance value.

$$\hat{j}_k \in \arg\min_{j=1,...,M+1} I(\sigma_j), \quad j = 1, 2, ..., M+1. \tag{5.28}$$

If more than one region is equally promising, the tie can be broken arbitrarily. If this index corresponds to a region that is a subregion of $\sigma(k)$, that is $\hat{j}_k \leq M$, then let this be the most promising region in the next iteration

$$\sigma(k+1) = \sigma_{\hat{j}_k}(k) \tag{5.29}$$

Otherwise, if the index corresponds to the complimentary region, that is $\hat{j}_k = M+1$, backtrack to the previous most promising region:

$$\sigma(k+1) = \sigma(k-1). \tag{5.30}$$

In the next section we evaluate the performance of this hybrid algorithm.

**Evaluation of the Hybrid NP/ACO Algorithm**

The hybrid NP/ACO algorithm described above is implemented by Al-Shihabi (2004) and tested on several hard TSP instances. The results of each instance are generated by running the algorithm 15 times using different random seeds. The parameters of the algorithm are set as follows: $\rho = 0.98$, $\alpha = 1.0$, $\beta = 2.0$, and $p_{best} = 0.05$.

The results are reported in Table 5.3, which reports the known optimum (column 2), the best, worst, and average solution found over the 15 replications of the NP/ACO hybrid (columns 3-5), and the average solution time (column 6). These numerical results indicate that the hybrid NP/ACO algorithm is capable of finding the global optimum solution for a number of instances and good quality results are obtained on the average for all of the TSP instances.

**Table 5.3.** Performance of Hybrid NP/ACO for TSP instances (see Table 1 in Al-Shihabi (2004).

| Problem | Optimum | NP/ACO Solution Best | Worst | Average | CPU Time |
|---|---|---|---|---|---|
| eil51 | 426 | 426 | 432 | 428 | 16 |
| berlin52 | 7542 | 7542 | 7762 | 7639 | 19 |
| eil75 | 538 | 538 | 544 | 541 | 50 |
| eil101 | 629 | 636 | 648 | 643 | 215 |
| krob150 | 26130 | 26257 | 28826 | 26527 | 1547 |
| d198 | 15780 | 15953 | 16129 | 16001 | 2538 |

## 5.3 Domain Knowledge in the NP Framework

In the previous two sections we have illustrated how to incorporate both general purpose metaheuristics and problem-specific local search methods into the NP framework to improve both the partitioning and the generation of feasible solutions. The third approach to improve the efficiency and effectiveness of the NP method is to incorporate expert domain knowledge in a similar manner, resulting in a *knowledge-based NP algorithm*.

For many complex, large-scale optimization problem in real systems there is great deal of functional expertise that if appropriately utilized could greatly improve the efficiency of the optimization. Many traditional methods cannot take advantage of such domain knowledge effectively, but the flexibility of the NP method allows for incorporating any such knowledge into either the partitioning or the generation of feasible solutions, leading to a knowledge-based NP implementation. The exact manner in which this is incorporated is application-specific, but here we outline some general strategies.

In many real problems it is known which variables are the most important in terms of their impact on the overall performance of the system. For example, in a scheduling problem there may be one or more known bottleneck(s) such that the schedule at the bottleneck(s) will largely determine the performance of the whole system. Such information can be utilized in the NP method by using it to define an intelligent partitioning. The overall goal of a good partitioning is to group together good solutions in the same region and poor solutions in other regions. This implies that the most important variables, in terms of their impact on the objective function, should be determined first. For example, variables corresponding to bottlenecks should be fixed at the top of the partitioning tree.

Another example of incorporation of domain knowledge is when partial solution can be obtained without applying an optimization algorithm. Consider for example the resource-constrained project scheduling problem introduced in section 1.3. The difficulty of this problem lies in the resource constraints, that is, each resource can only be used by one job or task at a time. However, in practice there may be a clear priority between some of the jobs. If an expert can specify such priorities, that is, generate a partial solution, then the generation of the remaining solution is much easier. The more priorities that can be specified using domain knowledge the easier the optimization problem becomes.

While some general guidelines and principles can be established for incorporating domain knowledge into the NP method, the details are always application-specific. A detailed example of how to accomplish this is given in Chapter 11, where we develop a knowledge-based NP algorithm for a process planning problem. The numerical results reported in that chapter show that incorporating domain knowledge results in significant improvements in algorithm efficiency and effectiveness.

# Part II

# Applications

# 6

# Flexible Resource Scheduling

This chapter illustrates how to take advantage of the flexibility in how complex problems can be formulated when solved using the NP method. Specifically, we consider a scheduling problem that is initially formulated as a binary integer program, which is consistent with the use of integer programming methods for solving the problem. Since the NP method only requires the ability to effectively partition and use random sampling to generate feasible solution, this allows for more flexible formulation of the problem. A reformulation of the problem can thus be obtained that reduces the feasible region by limiting it to all active schedules, and lends itself to a natural method of partitioning that enables imposing a structure on the feasible region. As noted in Section 2.2, such intelligent partitioning can greatly improve the efficiency of the NP method. Finally, this chapter also illustrates how both biased random sampling and greedy local search can be used to improve the generation of feasible solutions (see Section 2.3 and Section 5.1).

The application we consider is a scheduling problem that arises in many production systems. Production scheduling is often performed along several dimensions, including assigning jobs to manufacturing cells, sequencing the jobs within each cell, and allocating resources to the cells. Sometimes two or more of these scheduling decisions must be made simultaneously. An example of this could be an assembly system where several product families are assembled. If requirements vary drastically from one family to another, each cell may be uniquely configured to produce specific families and each family must therefore be processed in a given cell. On the other hand, workers may be cross-trained and could be dynamically assigned to any of the cells. The scheduling problem is therefore to simultaneously sequence the jobs within each cell, determine how many workers should be assigned to each cell at any given time, and find a starting time for each job.

Scheduling for parallel manufacturing systems has traditionally focused on two main issues: assigning jobs to machines and sequencing the jobs that are assigned to the same machine (Cheng and Sin 1990, Pinedo 1995). A few studies have focused on scheduling, either, job assignment or sequencing, when the speed of the machines is variable, such as is the case for many tooling machines (Adiri and Yehudai 1987, Trick 1994, Karabati and Kouvelis, 1997). On the other hand, in cellular manufacturing systems the processing speed may depend on some flexible resource, such as skilled labor, that is competed for by the cells. The problem of simultaneously allocating these flexible resources to the cells and scheduling the jobs within each cell has received relatively little attention, and these two problems have traditionally been considered separately. For example, Frenk *et al.* (1994) and Boxma *et al.* (1990) consider resource allocation in manufacturing systems, whereas So (1990) considers the problem of finding a feasible schedule for each machine in a parallel manufacturing system. Simultaneous job scheduling and resource allocation was first discussed by Daniels and Mazzola (1994). The two problems can be seen to be intimately related and considering them simultaneously may produce significant benefits (Karabati *et al.* 1995, Daniels *et al.* 1996). In related work, joint sequencing and resource allocation decisions have also been investigated for single machine shop scheduling (Vickson 1980, Van Wassenhove and Baker 1982, Daniels and Sarin 1989). However, in this context the resources are necessarily non-renewable and there is no competition for resources across manufacturing cells.

## 6.1 The PMSFR Problem

A cellular manufacturing system has a set of jobs, indexed by $i \in \mathcal{N} = \{1, 2, ..., n\}$, to be processed in one of $m$ manufacturing cells, indexed by $j \in \mathcal{M} = \{1, 2, ..., m\}$. There is a fixed amount $R$ of renewable resources that are flexible. The set $\mathcal{N}_j$ of jobs processed in cell $j \in \mathcal{M}$ is fixed, but the sequence of jobs within a cell, and the amount of flexible resources used in each cell at any time can be varied. The objective is to minimize the makespan $C_{max} = \max_{i \in \mathcal{N}} C_i$, where $C_i$ is the completion time of job $i \in \mathcal{N}$. This objective is often of interest for parallel systems, in particular since schedules with low makespan tend to balance the load between cells (Pinedo, 1995).

The optimization problem can be seen to be threefold. First, a dynamic resource allocation must be determined. Second, a sequence in which jobs should be processed within each cell must be determined, and third, a starting time for each job must be found. This scheduling problem has previously been discussed in Daniels *et al.* (1996) where it is termed the Parallel-Machine Flexible-Resource Scheduling (PMFRS) problem. The decision variables are defined as

$$y_{ih} = \begin{cases} 1 \text{ if job } i \text{ precedes job } h, \text{ where } i, h \in \mathcal{N}_j, \ j \in \mathcal{M}, \\ 0 \text{ otherwise.} \end{cases}$$

$$x_{irt} = \begin{cases} 1 \text{ if job } i \text{ completes processing with } r \text{ resources at time } t, \\ 0 \text{ otherwise.} \end{cases}$$

We let $\hat{p}_{ir}$ denote the service time of job $i \in \mathcal{N}$ when $r \leq R$ resources are allocated to the $i$th job and $p_i$ denote the actual service time of a job $i \in \mathcal{N}$. We also let $T$ be an upper bound on the makespan, for example $T = \max_{j \in \mathcal{M}} \sum_{i \in \mathcal{N}_j} \hat{p}_{i1}$. In terms of the decision variables $\mathbf{x} = \{x_{irt}\}$ and $\mathbf{y} = \{y_{ih}\}$, the PMFRS optimization problem can be formulated as the following binary integer programming (BIP) problem.

$$\min_{\mathbf{x},\mathbf{y}} C \tag{6.1}$$

s.t.

$$\sum_{r=1}^{R} \hat{p}_{ir} \sum_{t=1}^{T} x_{irt} = p_i, \ i \in \mathcal{N}, \tag{6.2}$$

$$\sum_{t=1}^{T} t \left( \sum_{r=1}^{R} x_{irt} \right) = C_i, \ i \in \mathcal{N}, \tag{6.3}$$

$$\sum_{r=1}^{R} \sum_{t=1}^{T} x_{irt} = 1, \ i \in \mathcal{N}, \tag{6.4}$$

$$C_i \leq C, \ i \in \mathcal{N}, \tag{6.5}$$

$$C_h - C_i + T(1 - y_{ih}) \geq p_h, \ i \in \mathcal{N}_j, \ h \in \mathcal{N}_j \setminus \{i\}, \ j \in \mathcal{M}, \tag{6.6}$$

$$y_{ih} + y_{hi} = 1, \ i \in \mathcal{N}_j, \ h \in \mathcal{N}_j \setminus \{1, 2, ..., i\}, \ j \in \mathcal{M}, \tag{6.7}$$

$$\sum_{i \in \mathcal{N}} \sum_{r=1}^{R} \sum_{l=t}^{t+\hat{p}_{ir}-1} r \cdot x_{irl} \leq R, \ t \in \{1, 2, ..., T\}, \tag{6.8}$$

$$y_{ih}, x_{irt} \in \{0, 1\}, \ 1 \leq r \leq R, \ i \in \mathcal{N}_j, \ h \in \mathcal{N}_j \setminus \{i\}, \ j \in M, \ t \in \{1, 2, ..., T\}, \tag{6.9}$$

$$C_i \geq 0, \ i \in \mathcal{N}. \tag{6.10}$$

Daniels *et al.* (1996) have shown how this problem can be solved using a branch-and-bound algorithm and have developed a heuristic that is based on its static equivalent. They show that the problem is NP-hard, and solve it exactly using the branch-and-bound method for relatively small problem instances, and for larger problems using the heuristic. This heuristic is based on the static equivalent of the PMFRS problem, where the resources are not flexible and stay in the same cell for the entire time horizon. It is readily seen that for the static problem the sequence of jobs does not matter and the problem reduces to a simple resource allocation problem. This problem can be solved efficiently to optimality (Daniels *et al.*, 1996), and is therefore a convenient benchmark for evaluating the performance of algorithms for the PMFRS problem, as well as for evaluating the benefits of flexible resources.

## 6.2 Reformulation of the PMSFR Problem

We define a vector $r = \{r_i\}_{i \in \mathcal{N}}$ to denote the resources allocated to each job and a vector $t = \{t_i\}_{i \in \mathcal{N}}$ to denote the starting time of each job. Specifying these two vectors completely determines a schedule $x = (R, T)$. In addition to being integer valued we assume the following constraints on $R \in \mathbf{N}^n$ and $T \in \mathbf{N}^n$. At any time no more than $R$ resources are allocated to the jobs currently being processed. Furthermore, for each job $i \in \mathcal{N}$, changing the starting time $t_i$ to $t_i - 1$ violates the resource restrictions. This means that each job is scheduled as early as possible without interfering with any other jobs. Schedules with this property are often referred to as active schedules (Pinedo, 2000). Lastly, only one job is processed at a time in each cell.

This reformulated PMFRS optimization problem can be stated mathematically as follows:

$$\min_{x=(R,T)} C_{max}, \tag{6.11}$$

s.t.

$$\sum_{i=1}^{n} r_i \cdot \chi_{\{t_i,...,t_i+\hat{p}_{ir_i}\}}(t) \leq R, \ t = 1, 2, ..., T, \tag{6.12}$$

$$\sum_{i=1}^{n} r_i \cdot \chi_{\{t_i,...,t_i+\hat{p}_{ir_i}\}}(t_{\tilde{i}} - 1) > R - r_{\tilde{i}}, \ \tilde{i} \in \mathcal{N}, \tag{6.13}$$

$$\sum_{i \in \mathcal{N}_j} \chi_{\{t_i,...,t_i+\hat{p}_{ir_i}\}}(t) \leq 1, \ t = 1, 2, ..., T, \ j \in \mathcal{M}, \tag{6.14}$$

$$t_i \leq C_{max}, i \in \mathcal{N}, \tag{6.15}$$

$$r, t \in \mathbf{N}^n. \tag{6.16}$$

Here $\chi_A(\cdot)$ is the indicator function, $\chi_A(t) = 1$ if $t \in A$ and $\chi_A(t) = 0$ if $t \notin A$. The constraint (6.12) ensures that the available amount of resources is not exceeded. The constraint (6.13) ensures that all schedules are active. Constraint (6.14) ensures that only one job in each cell is scheduled at the same time, and constraint (6.15) defines the makespan $C_{max}$, that is, the completion time of the last job. The feasible region is given by

$$X = \{(r, t) \in \mathbf{N}^n \times \mathbf{N}^n \mid \text{equations (6.12)-(6.14) hold}\},$$

and a schedule may be written as $\{(r_1, t_1), (r_2, t_2), ..., (r_n, t_n)\}$.

As an illustration, consider the simple example shown in Figure 6.1. This system has $m = 2$ manufacturing cells, $n = 3$ jobs, and $R = 2$ flexible resources. The schedule in Figure 6.1(a) is feasible. It allocates all of the flexible resources to each job while being processed and the two cells are therefore never in operation simultaneously. The schedule in Figure 6.1(b) divides the resource between the second job in the first cell and the job in the second cell. These jobs are therefore processed in parallel after the first job is completed.

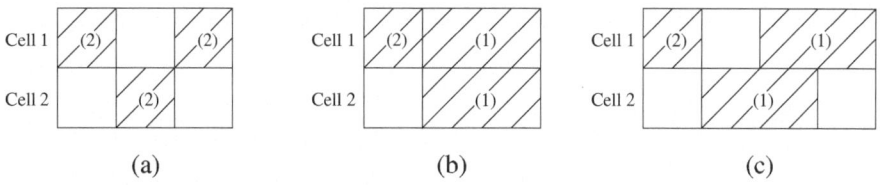

Cell 1 (2) (2)    Cell 1 (2) (1)    Cell 1 (2) (1)

Cell 2 (2)    Cell 2 (1)    Cell 2 (1)

(a)    (b)    (c)

**Fig. 6.1.** Possible schedules. Shaded areas indicate jobs. Number of resources is given in brackets

This schedule is also feasible. The schedule in Figure 6.1(c) is infeasible. This is because starting the second job of the first cell earlier would not violate the resource constraint, so this schedule violates constraint (6.13). We note that this schedule would be feasible according to the original formulation but that nothing is lost by making such schedules infeasible.

As illustrated in the above example the new formulation of the PMFRS reduces the feasible region. However, the optimal schedule(s) are retained as is established in the next theorem.

**Theorem 6.1.** *For any feasible solution to equations (6.13)-(6.16) there is a corresponding solution to equations (6.2)-(6.10) that has exactly the same makespan. Conversely, for any feasible solution to equations (6.2)-(6.10) there is a corresponding solution to equations (6.13)-(6.16) that has makespan that is less than or equal to the makespan of the first solution.*

*Proof:* (a) First assume that a schedule $(r, t)$ satisfies equations (6.13)-(6.16). Define the equivalent decision variables

$$y_{ih} = \begin{cases} 1 & t_i < t_h, \ i, h \in \mathcal{N}, \\ 0 & \text{otherwise.} \end{cases}$$

$$x_{irt} = \begin{cases} 1 & r = r_i, \ t = t_i + \hat{p}_{ir_i}, \ i \in \mathcal{N}, \\ 0 & \text{otherwise.} \end{cases}$$

First we must show that these decision variables are feasible for the original formulation, that is, that they satisfy the constraints (6.2) - (6.10). Note that for all $i \in \mathcal{N}$

$$\sum_{r=1}^{R} \hat{p}_{ik} \sum_{t=1}^{T} x_{irt} = \sum_{r=1}^{R} \hat{p}_{ik} \chi_{\{r_i\}}(r) = \hat{p}_{ir_i} = p_i.$$

Therefore equation (6.2) holds. Similarly, for all $i \in \mathcal{N}$

$$\sum_{t=1}^{T} t \left( \sum_{r=1}^{R} x_{irt} \right) = \sum_{t=1}^{T} t \chi_{\{t_i + \hat{p}_{ir_i}\}}(t) = t_i + \hat{p}_{ir_i} = C_i.$$

The constraints (6.4) and (6.5) hold trivially. Now let $j \in \mathcal{M}$, $i \in \mathcal{N}$, and $h \in \mathcal{N} \setminus \{i\}$. If $y_{ih} = 0$ then equation (6.6) is trivially satisfied. If $y_{ih} = 1$ then $t_i < t_h$ by definition and

$$C_h - C_i + T(1 - y_{ih}) = C_h - C_i = t_h + \hat{p}_{hr_h} - t_i + \hat{p}_{ir_i}$$
$$\geq \hat{p}_{hr_h} - \hat{p}_{ir_i} = p_h - p_i \geq p_h,$$

so the constraint (6.6) holds. Now for any cell $j \in \mathcal{M}$ and jobs in that cell $i \in \mathcal{N}$ and $h \in \mathcal{N} \setminus \{i\}$, then by equation (6.14), either $t_i < t_h$ or $t_i > t_h$. Hence either $y_{ih} = 1$ and $y_{hi} = 0$, or $y_{ih} = 0$ and $y_{hi} = 1$. Consequently $y_{ih} + y_{hi} = 1$ and equation (6.7) holds. Now for any $t \in \{1, 2, ..., T\}$

$$\sum_{i \in \mathcal{N}} \sum_{r=1}^{R} \sum_{l=t}^{t+\hat{p}_{ir}-1} r \cdot x_{irl} = \sum_{i \in \mathcal{N}} \sum_{l=t}^{t+\hat{p}_{ir}-1} \sum_{r=1}^{R} r \cdot x_{irl}$$

$$= \sum_{i \in \mathcal{N}} \sum_{l=t}^{t+\hat{p}_{ir}-1} r_i \chi_{\{t_i + \hat{p}_{ir_i}\}}(l)$$

$$= \sum_{i \in \mathcal{N}} r_i \sum_{l=t}^{t+\hat{p}_{ir}-1} \chi_{\{t_i + \hat{p}_{ir_i}\}}(l)$$

$$= \sum_{i \in \mathcal{N}} r_i \cdot \chi_{\{t_i, ..., t_i + \hat{p}_{ir_i}\}}(t)$$

$$\leq R.$$

This shows that constraint (6.8) holds. Equations (6.9) and (6.10) hold trivially. The feasibility of the solution has now been established. The fact that $(x, y)$ and $(r, t)$ have the same makespan is trivial.

(b) Now assume that we start with a feasible solution $(x, y)$ to the original problem. A feasible schedule $(r, t)$ that has a makespan less than or equal to the makespan corresponding to $(x, y)$ can be constructed as follows: Let the resource vector be defined as

$$r_i = \sum_{t=1}^{T} \sum_{r=1}^{R} r \cdot x_{irt}, \quad i \in \mathcal{N}. \tag{6.17}$$

Then the starting times can be obtained iteratively. Step 0. Let $j = 1$ and $\mathcal{N}_S = \emptyset$. Step 1. Let $\mathcal{N}_j^U = \mathcal{N}_j$ and $t_0 = 1$. Step 2. Select $i \in \{h \in \mathcal{N}_j^U : y_{hg} = 1, \forall g \in \mathcal{N}_j^U \setminus \{h\}\}$. Step 3. Let

$$t_i = \min \left\{ t \geq t_0 : R - \sum_{h \in \mathcal{N}_S} r_h \cdot \chi_{\{t_h, t_h+1, ..., t_h+\hat{p}_{hr_h}\}}(u) \geq r_i, u = t, ..., t + \hat{p}_{ir_i} \right\}.$$

Let $t_0 = t_i + \hat{p}_{ir_i} + 1$. Step 4. Let $\mathcal{N}_j^U = \mathcal{N}_j^U \setminus \{i\}$ and $\mathcal{N}_S = \mathcal{N}_S \cup \{i\}$. If $\mathcal{N}_j^U \neq \emptyset$ then go back to Step 2. Otherwise continue to Step 5. Step 5. If $j < m$, let $j = j + 1$ and go back to Step 1. Otherwise stop. It is clear that this construction maintains feasibility, that is, $(r, t)$ satisfies equations (6.12)-(6.16). Therefore, all that remains is to show that the makespan corresponding to $(r, t)$ is less than or equal to the makespan corresponding to the solution $(x, y)$ of the original problem, that is,

$$\max_{j \in \mathcal{M}} \max_{h \in \mathcal{N}_j} \left\{ \sum_{r=1}^{R} \sum_{t=1}^{T} t \cdot x_{hrt} \right\} \geq \max_{j \in \mathcal{M}} \max_{h \in \mathcal{N}_j} \left\{ t_h + \hat{p}_{hr_h} \right\}. \qquad (6.18)$$

We will prove this by contradiction. Assume that equation (6.18) does not hold. Then there exists at least one cell $j' \in \mathcal{M}$ such that

$$\max_{h \in \mathcal{N}_{j'}} \left\{ \sum_{r=1}^{R} \sum_{t=1}^{T} t \cdot x_{hrt} \right\} < \max_{h \in \mathcal{N}_{j'}} \left\{ t_h + \hat{p}_{hr_h} \right\}.$$

Furthermore, since the construction algorithm preserves the original sequencing, these maxima are realized for the same job $i' \in \mathcal{N}_{j'}$

$$\sum_{r=1}^{R} \sum_{t=1}^{T} t \cdot x_{i'rt} < t_{i'} + \hat{p}_{i'r_{i'}}.$$

However, since $(x, y)$ is feasible then $t_{i'}^* = \sum_{r=1}^{R} \sum_{t=1}^{T} t \cdot x_{i'rt} - \hat{p}_{i'r_{i'}}$ is a feasible starting time for job $i' \in \mathcal{N}_{j'}$ and $t_{i'}^* < t_{i'}$. This violates constraint (6.13) and is therefore a contradiction.

This theorem shows that the optimal solution to the PMFRS problem is an active schedule, so all the optimal schedules are contained in the reduced feasible region defined by equations (6.11) - (6.16) above.

## 6.3 NP Algorithm for the PMSFR Problem

We now describe how the NP method can be implemented for the reformulated PMFRS problem defined by equations (6.11) - (6.16) above. This implementation takes advantage of the special structure of the PMFRS problem both in defining an intelligent partitioning and in generating good feasible solutions from each region.

### 6.3.1 Partitioning

First we address the partitioning. The basic idea is to completely schedule one job at each level of the partitioning tree so that a region of depth $d$ is defined by $d$ jobs being scheduled, or fixed, and by the remaining $n - d$ jobs being free. By equation (6.13) each job is scheduled as early as possible given all other jobs being fixed. Accordingly, we adopt the rule of scheduling each job as early as possible given all the jobs that have already been fixed. Therefore, selecting a job to be fixed early gives the job certain priority. In other words, jobs that are assigned resources and starting time at low depth tend to be scheduled earlier than jobs that are fixed at greater depth. This imposes a structure on the feasible region and allows for special structure to be incorporated into the manner in which good solutions are generated.

Recall that a schedule can be represented as

$$(r, t) = \{(r_1, t_1), (r_2, t_2), ..., (r_n, t_n)\},$$

and the partitioning fixes one of these pairs $(r_{i_d}, t_{i_d})$ to a given value at depth $d$ in the partitioning tree, $d = 1, 2, ..., n$. The depth one subregions are therefore defined by the following $R \cdot n$ sets:

$$\{ (r, 1), (\cdot, \cdot), ..., (\cdot, \cdot) \}, r \in \{1, 2, ..., R\},$$
$$\{ (\cdot, \cdot), (r, 1), ..., (\cdot, \cdot) \}, r \in \{1, 2, ..., R\},$$
$$\vdots \qquad\qquad \vdots$$
$$\{ (\cdot, \cdot), (\cdot, \cdot), ..., (r, 1) \}, r \in \{1, 2, ..., R\}.$$

The depth-two regions are similarly defined by fixing two elements and so forth.

Given a region $\sigma \in \Sigma$ that determines $d(\sigma)$ of these pairs, we let $r(\sigma)$ denote the levels of resources and $t(\sigma)$ the starting times for the $d(\sigma)$ jobs that have already been scheduled. To describe the partitioning procedure in detail it is convenient to establish additional notation. We let $\mathcal{N}_j^U(\sigma)$ denote the set of unscheduled jobs in cell $j \in \mathcal{M}$ and $\mathcal{N}^U(\sigma) = \cup_{j \in \mathcal{M}} \mathcal{N}_j^U(\sigma)$ denote all the unscheduled jobs. We let $t_j^0(\sigma)$ denote the first possible starting time in cell $j$,

$$t_j^0(\sigma) = \max_{i \in \mathcal{N}_j \setminus \mathcal{N}_j^U(\sigma)} \left\{ t_i(\sigma) + \hat{p}_{ir_i(\sigma)} + 1 \right\}.$$

Finally, we let the vector $r^A(\sigma) = \{r_t^A(\sigma)\}_{t=1,2,...,T}$ denote the available resources, i.e., the resources that have not been allocated, at any given time. In terms of $(r, t)$ it can be written as

$$r_t^A(\sigma) = R - \sum_{i=1}^{n} r_i(\sigma) \cdot \chi_{\{t_i, ..., t_i + \hat{p}_{ir_i(\sigma)}\}}(t).$$

We can now describe the partitioning as follows: Assume we want to partition a region $\sigma(k) \in \Sigma$. We let each subregion correspond to scheduling one of the jobs in $\mathcal{N}^U(\sigma(k))$ next with a given amount of resources. Each subregion is otherwise identical to $\sigma(k)$. Assume that job $i \in \mathcal{N}_j^U(\sigma(k))$ determines a subregion $\sigma_l(k)$ with $r \in \{1, 2, ..., R\}$ resources allocated to this job. Then calculate its starting time as the earliest feasible starting time

$$t_i(\sigma_l(k)) = \arg \min_{t \geq t_j^0(\sigma(k))} \{r_t^A(\sigma(k)) \geq r\}. \tag{6.19}$$

Generate a processing time $p_i = \hat{p}_{ir}$ for the job, and update the earliest feasible starting time in the cell for the subregion $t_j^0(\sigma_l(k)) = t_i(\sigma_l(k)) + p_i + 1$.

The earliest feasible starting time for all other cells is the same for $\sigma_l(k)$ as $\sigma(k)$. Finally, reduce the amount of available resources for the time job $i$ is being processed, that is, let $r_t^A(\sigma_l(k)) = r_t^A(\sigma_l(k)) - r$ for all $t$ such that $t_i(\sigma_l(k)) \leq t \leq t_i(\sigma_l(k)) + p_i$. This completely defines the subregion.

Figure 6.2 partially illustrates the partitioning tree for the simple sample problem discussed above. Recall that this example has $n = 3$ jobs to be processed on one of $m = 2$ cells which have $R = 2$ flexible resources at their disposal. At depth one there are

$$R \cdot \sum_{j \in \mathcal{M}} |N_j^U| = 2 \cdot (2 + 1) = 6$$

subregions, two of which are shown in Figure 6.2. We assume that the jobs to be processed in cell one are labelled as jobs one and two, and that the job in cell two is labelled as job three. We assume that all the jobs are identical with a processing time of one time unit if they have two resources, and two time units if they have only one resource. The six depth-one subregions can therefore be represented as

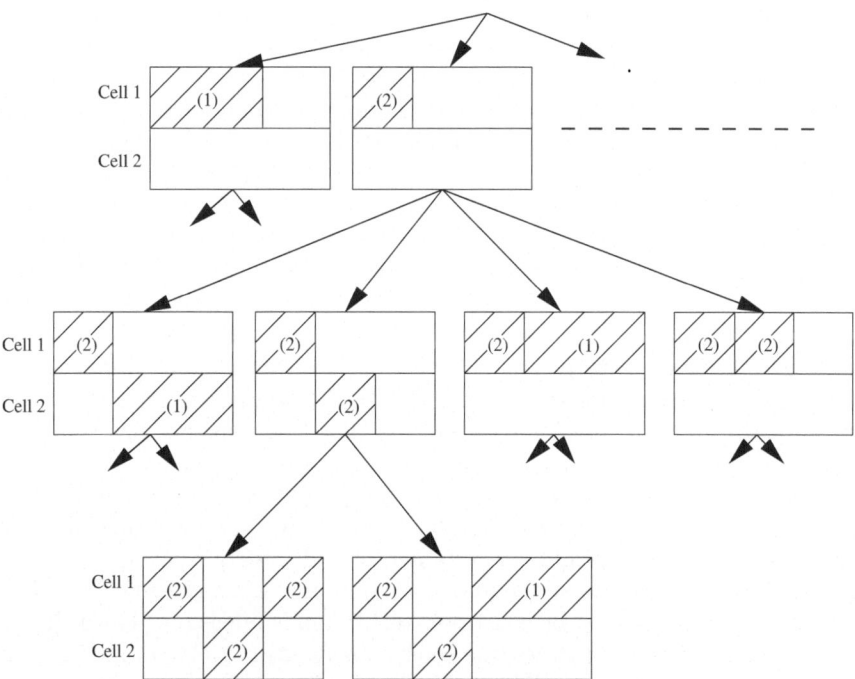

**Fig. 6.2.** Part of a partitioning tree.

$$\sigma_1(0) = \{(1,1), (\cdot, \cdot), (\cdot, \cdot)\}, \ \sigma_2(0) = \{(2,1), (\cdot, \cdot), (\cdot, \cdot)\},$$
$$\sigma_3(0) = \{(\cdot, \cdot), (1,1), (\cdot, \cdot)\}, \ \sigma_4(0) = \{(\cdot, \cdot), (2,1), (\cdot, \cdot)\},$$
$$\sigma_5(0) = \{(\cdot, \cdot), (\cdot, \cdot), (1,1)\}, \ \sigma_6(0) = \{(\cdot, \cdot), (\cdot, \cdot), (2,1)\}.$$

Now if in the next iteration $\sigma(1) = \sigma_1(0) = \{(2,1), (\cdot, \cdot), (\cdot, \cdot)\}$ then there are four subregions

$$\sigma_1(1) = \{(2,1), (\cdot, \cdot), (1,2)\}, \ \sigma_2(1) = \{(2,1), (\cdot, \cdot), (2,2)\},$$
$$\sigma_3(1) = \{(2,1), (1,2), (\cdot, \cdot)\}, \ \sigma_4(1) = \{(2,1), (2,2), (\cdot, \cdot)\}.$$

Now if, as is illustrated in Figure 6.2, the next most promising region is $\sigma(2) = \{(2,1), (\cdot, \cdot), (2,2)\}$ then there are only two subregions

$$\sigma_1(2) = \{(2,1), (1,3), (2,2)\}, \ \sigma_2(2) = \{(2,1), (2,3), (2,2)\}.$$

These subregions are now singletons and completely define a feasible schedule.

The next proposition establishes that this is in fact a valid method of partitioning.

**Theorem 6.2.** *Every feasible schedule, i.e. point in $X$, corresponds to a maximum depth region obtained by applying the above partitioning procedure. Conversely, every maximum depth region corresponds to a point in $X$.*

*Proof:* Let $x \in X$ and write $x = (r, t)$. We need to identify a maximum depth region that corresponds to this point. The partitioning procedure sequentially selects the jobs and schedules them at the first feasible point. Hence, given $x$, we need to construct a sequence of jobs $i_1, i_2, ..., i_n$ with the following property. Assume that job $i_l$ is given resources $r_{i_l}$, $l = 1, 2, ..., n$. If jobs $i_1, .., i_l$ have already been scheduled, then the earliest feasible time for job $i_{l+1}$ to start is $t_{i_{l+1}}$, $l = 1, 2, ..., n - 1$. We can generate this sequence as follows: Initialize by setting the list of scheduled jobs to be empty, $\mathcal{N}_s = \emptyset$. In the first iteration, select a job that has the first possible starting time, i.e. select $i_1 \in \arg\min_{i \in \mathcal{N} \backslash \mathcal{N}_s} t_i$, and let $\mathcal{N}_s = \mathcal{N}_s \cup \{i_1\}$. If more than one are started at the same time, one may be selected arbitrarily. Now simply repeat the process, i.e. select a job $i_2 \in \arg\min_{i \in \mathcal{N} \backslash \mathcal{N}_s} t_i$. This is repeated until there are no more jobs. We use induction to see that this sequence has the right properties. First consider job $i_1$. The earliest feasible starting time is $t = 0$. But since $x$ is a feasible point, we must also have $t_{i_1} = 0$ because otherwise $t_i > 0$ for all $i \in \mathcal{N}$ and constraint (6.13) is violated. Now assume that the jobs $i_1, ..., i_{l-1}$ have the right property and consider the job $i_l$. Let $t_{i_l}$ be the first feasible time given that the jobs $i_1, ..., i_{l-1}$ have already been scheduled. If $t_{i_l} < t_{i_l}$ then $x$ violates the constraint (6.13) and is infeasible. This is hence a contradiction. If $t_{i_l} > t_{i_l}$ then the constraint (6.12) is violated and again $x$ is infeasible. Hence $t_{i_l} = t_{i_l}$. This holds for any $l \in \{2, 3, ..., n\}$, so the entire sequence has the desired property.

Conversely, since the partitioning procedure never exceeds the available resources and each job is scheduled as soon as the resources allow, then every maximum depth region corresponds to a feasible schedule.

We have now established a valid partitioning procedure. The next step is to develop an efficient procedure to randomly sample a region.

### 6.3.2 Generating Feasible Solutions

Recall that the only constraint on how feasible solutions should be generated from each region is that each schedule should have a positive probability of being selected. When a region $\sigma \in \Sigma$ is sampled then $d(\sigma)$ jobs have already been assigned a fixed amount of resources and a starting time. For each of the $n - d(\sigma)$ jobs that have not been fixed the sampling procedure involves selecting cells, jobs, and resources. Sampling can always be done in a generic fashion, for example uniformly, but in practice it may be better to attempt to bias the sampling distribution towards good solutions (see Section 2.3.1). Since jobs selected to be fixed early tend to be scheduled earlier, and it is desirable to balance the workload, it is intuitive that higher priory should be given to selecting jobs in cells that have much work waiting. Also, there is an intuitive appeal to selecting with high probability resource levels that enable immediate scheduling of a job. This motivates the versatile sampling algorithm given below. The intuition behind this algorithm is that since the objective is to minimize the makespan it may be best to finish quickly as much work as possible in the cells that have the most work waiting. Thus, priority is given to the most heavily-loaded cells and long jobs in those cells.

Assume that a region $\sigma \in \Sigma$ is being sampled. Then the starting time and number of allocated resources have already been determined for the first $d(\sigma)$ jobs. The sampling algorithm fixes the remaining $n - d(\sigma)$ jobs iteratively and generates a single sample schedule $(\hat{r}, \hat{t})$.

### Sampling Algorithm

Step 0.    Initialize $\mathcal{N}_j^U = \mathcal{N}_j^U(\sigma)$, $j \in \mathcal{M}$.

Repeat until $\mathcal{N}_j^U = \emptyset$ for all $j \in \mathcal{M}$.

Step 1.    Determine the cell $\hat{J}$ from which the next job is selected. Higher priority is given to cells that have much work waiting to be processed, so $\hat{J}$ is a random variable with

$$
P\left[\hat{J}=j\right] = \begin{cases} \beta_1 \cdot \dfrac{\sum\limits_{i\in\mathcal{N}_j^U}\hat{p}_{i1}}{\sum\limits_{k\in\mathcal{M}}\sum\limits_{i\in\mathcal{N}_k^U}\hat{p}_{i1}} + \sum\limits_{k\in\mathcal{M}}\dfrac{1-\beta_1}{\chi_{\mathcal{N}_k^U\neq\emptyset}(k)}, & \mathcal{N}_j^U\neq\emptyset, \\[20pt] 0 & \text{otherwise.} \end{cases}
$$
(6.20)

Here $\beta_1$ is a constant to be determined.

Step 2. Given an outcome $\hat{j}\in\mathcal{M}$ select a job $\hat{I}$ within this cell. Higher priority is given to long jobs, so $\hat{I}$ is a random variable with

$$
P\left[\hat{I}=i\right] = \begin{cases} \beta_2\cdot\dfrac{\hat{p}_{i1}}{\sum\limits_{k\in\mathcal{N}_{\hat{j}}^U}\hat{p}_{k1}} + (1-\beta_2)\cdot\dfrac{1}{|\mathcal{N}_{\hat{j}}^U|}, & i\in\mathcal{N}_{\hat{j}}^U, \\[20pt] 0 & \text{otherwise.} \end{cases}
$$
(6.21)

Here $\beta_2$ is a constant to be determined.

Step 3. Given a selected job $\hat{i}\in\mathcal{N}$ select the number of resources $\hat{r}$ given to this job. Higher priority is given to resource levels that allow for immediate scheduling, that is, resource levels that are less than or equal to $r^A_{t^0_{\hat{j}}}$, the available resources at the the earliest feasible time $t^0_{\hat{j}}$ in the cell $\hat{j}\in\mathcal{M}$. Then $\hat{r}$ is a random variable with

$$
P\left[\hat{r}_{\hat{i}}=r\right] = \begin{cases} \beta_3\cdot\dfrac{\beta_4}{r^A_{t^0_{\hat{j}}}} + (1-\beta_3)\cdot\dfrac{1}{R}, & r\le r^A_{t^0_{\hat{j}}}, \\[20pt] \beta_3\cdot\dfrac{1-\beta_4}{R-r^A_{t^0_{\hat{j}}}} + (1-\beta_3)\cdot\dfrac{1}{R}, & r> r^A_{t^0_{\hat{j}}}. \end{cases}
$$
(6.22)

Here $\beta_3$ and $\beta_4$ are constants to be determined. A weight $\beta_4$ is given to jobs that can be scheduled immediately and as before this biased sampling is weighted with uniform sampling.

Step 4. Schedule job $\hat{i}\in\mathcal{N}_{\hat{j}}$ using $\hat{r}$ resources at the first feasible time $\hat{T}_{\hat{i}}$.

Step 5. If $\mathcal{N}_j^U=\emptyset$ for all $j\in\mathcal{M}$ then stop. Otherwise go back to Step 1.

If $\beta_1=\beta_2=\beta_3=0$ then this is a uniform sampling procedure. If these constants are positive, priority is given to cells that have many jobs waiting, long jobs, and resource allocations that enable immediate scheduling of the selected job. In particular, we refer to $\beta_1=\beta_2=\beta_3=1$ as weighted sampling. If $\beta_3>0$ then $\beta_4$ determines how much priority is given to resource levels that allow for immediate scheduling of the selected job.

The biased sampling procedure generates higher-quality solutions from each region, but this can be improved further by incorporating a heuristic search into the generation procedure (see Section 2.3.2 and Section 5.1). We let $\mathcal{D}(\sigma)$ denote the set of schedules sampled from region $\sigma\in\Sigma$. If we have at our disposal a heuristic $H_\sigma:X\to X$ that transforms an initial schedule $x^0\in\sigma$ into an improved schedule $x^1\in\sigma$, this can be incorporated into the solution generation. Thus, the new set $\left\{H_\sigma(x^0):x^0\in\mathcal{D}(\sigma)\right\}$ is used as the basis for selecting the next most promising region. The improvement heuristic can, for example, be the efficient heuristic proposed in Daniels *et al.* (1996). However, the efficiency of the NP algorithm also depends on the random sampling algorithm, the structure imposed by the partitioning , and repeated use of the improvement heuristic. It may therefore be preferable to have a simpler heuristic that can be repeatedly called with relatively little computational cost. For example, we can seek simple changes to improve the resource allocation. Such improvements should be made by allocating more resources to critical jobs, that is, jobs which are such that a delay in their starting time causes an increase in the makespan of the cell. This is the basic idea behind the improvement algorithm below.

Assume that a sample point $\{(r_{i_1}, t_{i_1}), ..., (r_{i_n}, t_{i_n})\}$ has been generated from $\sigma \in \Sigma$ using the random sampling procedure. Let $\mathcal{J}$ denote the set of jobs that have been modified.

## Improvement Algorithm

Step 0.   Initialize $\mathcal{J} = \emptyset$.

Repeat until $\mathcal{N}_j^U(\sigma) \setminus \mathcal{J} = \emptyset$ for all $j \in \mathcal{M}$.

Step 1.   Select a cell $\hat{j} \in \mathcal{M}$ such that $\mathcal{N}_{\hat{j}}^U(\sigma) \setminus \mathcal{J} \neq \emptyset$.

Step 2.   For each $i \in \mathcal{N}_{\hat{j}}^U(\sigma)$ let $\tilde{x}_i = \{(r_{i_1}, t_{i_1}), ..., (r_i, t_i + 1), ..., (r_{i_n}, t_{i_n})\}$, and calculate the set of critical jobs

$$\mathcal{C} = \left\{ i \in \mathcal{N}_{\hat{j}}^U(\sigma) : C_{\hat{j}}(\tilde{x}_i) - C_{\hat{j}}(x) > 0 \right\}. \qquad (6.23)$$

If $\mathcal{C} = \emptyset$ let $\mathcal{J} = \mathcal{J} \cup \mathcal{N}_{\hat{j}}^U(\sigma)$ and go back to Step 1 above. Otherwise select a job $\hat{i} \in \mathcal{C}$, let $\mathcal{J} = \mathcal{J} \cup \{\hat{i}\}$, and continue to Step 3 below.

Step 3.   If there are additional resources available during the time at which job $\hat{i}$ is being processed, that is, if $\tilde{R} = R - \sum_{i=1}^n r_i \cdot \chi_{\{t_i, ..., t_i + \hat{p}_{ir_i}\}}(t_i) > 0$, then there is a possibility of improvement. If such a possibility exists continue to Step 4. Otherwise, if $\mathcal{C} \cap (\mathcal{N} \setminus \mathcal{J}) \neq \emptyset$, go back to Step 2. Otherwise go back to Step 1.

Step 4.   If increasing the resources to job $\tilde{i}$ decreases the makespan of the cell then give it all available extra resources. That is, consider

$$\tilde{x} = \left\{ (r_{i_1}, t_{i_1}), ..., (r_{\tilde{i}} + \tilde{R}, t_{\tilde{i}}), ..., (r_{i_n}, t_{i_n}) \right\},$$

and if $C_j(\tilde{x}) < C_j(x)$ then let $x = \tilde{x}$.

Step 5.   If $\mathcal{C} \cap (\mathcal{N} \setminus \mathcal{J}) \neq \emptyset$, go to Step 2. Otherwise go to Step 1.

For this to be a valid heuristic to use to generate feasible solutions we must check that when it is applied to a schedule in a given region $\sigma \in \Sigma$ then the improved schedule is also in the region, and in particular that it agrees with the original performance measure on singleton regions.

**Theorem 6.3.** *Let $H_\sigma : X \to X$ denote the improvement algorithm described above.*
*(a) For all $\sigma \in \Sigma$,*

$$H_\sigma(\sigma) \subseteq \sigma. \tag{6.24}$$

*In particular if $\sigma = \{x\}$ is a singleton region then $H_\sigma(x) = x$.*
    *(b) For any $x \in X$ such that $H_\sigma(x) \neq x$,*

$$C\left(H_\sigma(x)\right) \leq C(x). \tag{6.25}$$

*Proof:* (a) Equation (6.24) follows directly from the observation that when applied to a region $\sigma \in \Sigma$ the improvement algorithm only makes changes to the jobs in $\mathcal{N}^U(\sigma)$. Thus, when $\sigma = \{x\}$ is a singleton region then $H_\sigma(x) = x$.
(b) Equation (6.25) holds because by Step 4 of the algorithm, changes in the schedule are only made if it decreases the makespan.

## 6.4 Numerical Example

We now illustrate the NP method for the reformulated PMFRS problem with a few numerical examples. The implementation of the NP algorithm is as follows: The sampling algorithm is chosen to be uniform $(\beta_1 = \beta_2 = \beta_3 = 0)$ in the subregions, and weighted $(\beta_1 = \beta_2 = \beta_3 = 1)$ in the complimentary region with weight $\beta_4 = \frac{1}{2}$. One sample schedule is constructed in each subregion, and ten in the complimentary region. This increased computational effort in the complimentary region helps enable the algorithm to backtrack when needed. In all regions the promising index is estimated by applying the improvement heuristic to the sample schedule(s). The processing times are generated in the following manner. For a job $i \in \mathcal{N}$, we let $\hat{p}_{i1}$ be generated from a uniform random variable $\mathcal{U}(10, 50)$, and the remaining processing times calculated according to

$$\hat{p}_{ir} = \left(1 - \alpha\left(1 - \frac{1}{k}\right)\right)\hat{p}_{i1},$$

where $\alpha$ is a constant that determines the speedup achieved when additional resources are available.

We start by considering how many iterations of the algorithm are required for obtaining near-optimal schedules. To that end we look at a problem with settings $n \in \{10, 20\}$, $R = 4$, $m \in \{3, 4, 5\}$, and $\alpha \in \{0.2, 0.5, 0.8\}$, for 100, 500, and 1000 iterations. The algorithm is repeated 100 times for each setting, so the averages are based on 100 independent replications. The results can be found in Table 6.1, which shows both the average and best makespan

**Table 6.1.** Performance of the NP algorithm for a variable number of iterations.

| Problem Settings | | | | 100 Iterations | | 500 Iterations | | 1000 Iterations | |
|---|---|---|---|---|---|---|---|---|---|
| $n$ | $R$ | $\alpha$ | $m$ | Average | Best | Average | Best | Average | Best |
| 10 | 4 | 0.2 | 3 | 101.8 | 99 | 100.8 | 98 | 100.6 | 98 |
| | | | 4 | 86.3 | 84 | 86.0 | 82 | 86.3 | 84 |
| | | | 5 | 79.2 | 75 | 78.2 | 75 | 77.5 | 74 |
| | | 0.5 | 3 | 92.9 | 87 | 91.3 | 87 | 91.4 | 88 |
| | | | 4 | 81.2 | 76 | 80.4 | 76 | 80.1 | 76 |
| | | | 5 | 76.4 | 70 | 75.6 | 70 | 74.9 | 70 |
| | | 0.8 | 3 | 79.7 | 75 | 78.4 | 75 | 77.8 | 74 |
| | | | 4 | 74.1 | 71 | 72.9 | 71 | 72.5 | 70 |
| | | | 5 | 72.0 | 69 | 71.1 | 69 | 70.9 | 68 |
| 20 | 4 | 0.2 | 3 | 193.0 | 184 | 193.0 | 186 | 191.3 | 181 |
| | | | 4 | 172.3 | 161 | 170.6 | 162 | 170.9 | 158 |
| | | | 5 | 162.0 | 137 | 160.7 | 133 | 159.9 | 154 |
| | | 0.5 | 3 | 179.9 | 167 | 178.7 | 169 | 178.9 | 166 |
| | | | 4 | 163.1 | 148 | 162.6 | 155 | 161.1 | 152 |
| | | | 5 | 154.9 | 148 | 153.3 | 145 | 152.7 | 137 |
| | | 0.8 | 3 | 160.3 | 147 | 159.6 | 149 | 158.9 | 152 |
| | | | 4 | 150.5 | 146 | 149.1 | 143 | 149.1 | 145 |
| | | | 5 | 144.1 | 138 | 142.9 | 138 | 142.6 | 138 |

found for these settings. From the data we see that substantial improvements in the average makespan are obtained by increasing the number of iterations from 100 to 500, but beyond that relatively little improvements are made for these examples. Furthermore, we note that for several settings the best makespan across all the replications is found when only 100 iterations are used, indicating that instead of using very long runs of the algorithm it may be more beneficial to use independent replications. Hence for the following numerical experiments we choose to use a moderate number of 500 iterations.

Since optimal schedules for the PMFRS problem cannot be found efficiently and the problems here are too large for complete enumeration, we evaluate the quality of schedules found by the NP algorithm by comparing the makespan of those schedules with the makespan of the optimal schedules for the static problem. Recall that for this problem only the resource allocation is of consequence, and an optimal schedule can be obtained efficiently. Since the makespan of the schedules obtained by the NP algorithm are also upper bounds for the optimal makespan of the PMFRS problem, this comparison also provides a lower bound on the performance improvement that may be achieved by making static resources flexible.

We run the NP method using the above settings for problems with $n \in \{10, 15\}$ jobs, $m \in \{3, 4, 5\}$ cells, $R \in \{4, 5, 6\}$ flexible resources, and a speedup factor of $\alpha \in \{0.2, 0.5, 0.8\}$. For each of these settings we use

**Table 6.2.** Performance of the NP algorithm for $n = 10$ after 500 iterations.

| Problem Settings | | | Makespan Found by NP | | | Static Problem | | Average |
|---|---|---|---|---|---|---|---|---|
| $n$ | $R$ | $\alpha$ | $m$ | Average | Min | S.E. | Makespan | % Over | CPU Time |
| 10 | 4 | 0.2 | 3 | 100.8 | 98 | 1.8 | 100 | 2.0% | 22.6 |
| | | | 4 | 86.0 | 82 | 2.6 | 91 | 11.0% | 16.5 |
| | | | 5 | 78.2 | 75 | 7.0 | † | | 19.2 |
| | | 0.5 | 3 | 91.3 | 87 | 2.7 | 100 | 14.9% | 22.5 |
| | | | 4 | 80.4 | 76 | 6.2 | 91 | 19.7% | 21.3 |
| | | | 5 | 75.6 | 70 | 6.0 | † | | 23.6 |
| | | 0.8 | 3 | 78.4 | 75 | 3.3 | 100 | 33.3% | 12.4 |
| | | | 4 | 72.9 | 71 | 3.6 | 91 | 28.2% | 10.8 |
| | | | 5 | 71.1 | 69 | 1.9 | † | | 15.6 |
| | 5 | 0.2 | 3 | 98.6 | 96 | 1.8 | 99 | 3.1% | 14.0 |
| | | | 4 | 82.1 | 79 | 5.4 | 80 | 1.3% | 35.3 |
| | | | 5 | 74.5 | 70 | 3.9 | 79 | 12.9% | 23.9 |
| | | 0.5 | 3 | 84.9 | 80 | 5.2 | 84 | 5.0% | 24.1 |
| | | | 4 | 72.9 | 68 | 7.9 | 71 | 4.4% | 21.1 |
| | | | 5 | 68.7 | 65 | 5.3 | 79 | 21.5% | 15.0 |
| | | 0.8 | 3 | 68.6 | 64 | 3.1 | 67 | 4.7% | 15.9 |
| | | | 4 | 63.4 | 61 | 1.7 | 71 | 16.4% | 17.8 |
| | | | 5 | 61.3 | 59 | 3.0 | 79 | 33.9% | 25.0 |
| | 6 | 0.2 | 3 | 96.5 | 95 | 1.1 | 96 | 1.1% | 32.2 |
| | | | 4 | 79.4 | 77 | 4.0 | 77 | 0.0% | 18.8 |
| | | | 5 | 71.7 | 67 | 4.2 | 70 | 4.5% | 20.3 |
| | | 0.5 | 3 | 79.6 | 74 | 4.3 | 75 | 1.4% | 12.7 |
| | | | 4 | 67.6 | 64 | 5.4 | 68 | 6.3% | 22.2 |
| | | | 5 | 63.1 | 58 | 6.2 | 63 | 8.6% | 25.2 |
| | | 0.8 | 3 | 61.7 | 58 | 2.9 | 60 | 3.4% | 34.4 |
| | | | 4 | 57.3 | 52 | 2.7 | 63 | 21.2% | 18.2 |
| | | | 5 | 53.6 | 49 | 3.0 | 63 | 28.6% | 23.5 |

† Problem setting infeasible for static problem.

100 replications. The results are reported in Table 6.2 - Table 6.3, which show the average and best makespan across the replications, as well as the standard error. The average CPU time in seconds is also reported. Since a fixed number of iterations is used, this does not vary much from one setting to another. Finally, as stated above, we report the makespan of the optimal static schedule for comparison, as well as the percentage improvement of the best PMFRS schedule found by the NP algorithm. The schedules found by the NP algorithm tend to have much better performance than the static schedules, which indicates that considerable performance benefits may be obtained through the use of flexible resources. As is to be expected, the performance improvements increase with the speedup factor. Furthermore, the performance improvements also increase with the number of cells ($m$), indicating that flex-

**Table 6.3.** Performance of the NP algorithm for $n = 15$ after 500 iterations.

| \multicolumn{3}{c}{Problem Settings} | | | \multicolumn{3}{c}{Makespan Found by NP} | | | \multicolumn{2}{c}{Static Problem} | | Average |
| $n$ $R$ $\alpha$ | | | $m$ | Average | Min | S.E. | Makespan | % Over | CPU Time |
|---|---|---|---|---|---|---|---|---|---|
| 15 4 0.2 | | | 3 | 148.6 | 139 | 12.7 | 139 | 0.0% | 23.7 |
| | | | 4 | 130.0 | 125 | 27.3 | 131 | 4.8% | 35.8 |
| | | | 5 | 123.7 | 109 | 60.6 | † | | 21.8 |
| | | 0.5 | 3 | 137.4 | 127 | 15.4 | 127 | 0.0% | 28.5 |
| | | | 4 | 124.7 | 115 | 18.6 | 131 | 13.9% | 33.3 |
| | | | 5 | 118.6 | 109 | 29.4 | † | | 54.8 |
| | | 0.8 | 3 | 122.7 | 114 | 7.1 | 127 | 11.4% | 33.3 |
| | | | 4 | 115.6 | 110 | 6.9 | 131 | 19.1% | 20.8 |
| | | | 5 | 111.6 | 107 | 9.8 | † | | 26.8 |
| | 5 0.2 | | 3 | 143.2 | 134 | 10.8 | 134 | 0.0% | 25.8 |
| | | | 4 | 123.4 | 115 | 10.8 | 115 | 0.0% | 32.0 |
| | | | 5 | 111.5 | 101 | 49.0 | 106 | 5.0% | 44.5 |
| | | 0.5 | 3 | 126.7 | 117 | 18.5 | 126 | 7.7% | 31.7 |
| | | | 4 | 114.1 | 103 | 20.3 | 114 | 10.7% | 23.2 |
| | | | 5 | 105.0 | 100 | 19.2 | 106 | 6.0% | 44.4 |
| | | 0.8 | 3 | 108.0 | 102 | 6.8 | 126 | 23.5% | 41.9 |
| | | | 4 | 101.6 | 96 | 7.2 | 114 | 18.8% | 31.6 |
| | | | 5 | 96.4 | 89 | 6.8 | 106 | 19.1% | 20.8 |
| | 6 0.2 | | 3 | 139.8 | 132 | 11.7 | 131 | 1.1% | 32.7 |
| | | | 4 | 119.8 | 112 | 12.8 | 114 | 1.8% | 39.2 |
| | | | 5 | 104.6 | 98 | 21.5 | 101 | 4.5% | 46.6 |
| | | 0.5 | 3 | 120.8 | 108 | 17.9 | 116 | 7.4% | 22.8 |
| | | | 4 | 107.9 | 98 | 17.7 | 101 | 3.1% | 32.3 |
| | | | 5 | 98.2 | 90 | 14.3 | 101 | 12.2% | 61.9 |
| | | 0.8 | 3 | 98.0 | 91 | 6.0 | 93 | 3.3% | 54.4 |
| | | | 4 | 91.1 | 86 | 5.8 | 101 | 17.4% | 25.0 |
| | | | 5 | 86.5 | 81 | 7.0 | 101 | 24.7% | 49.0 |

† Problem setting infeasible for static problem.

ible resources are most beneficial for systems where the speedup factor is large
and there is a large number of parallel cells.

## 6.5 Conclusions

In this chapter we have shown how the NP method can effectively address
the problem of simultaneously obtaining a resource allocation and sequencing jobs in a cellular manufacturing system with parallel cells and flexible resources. Specifically, we solved the Parallel-Machine Flexible-Resource
Scheduling (PMFRS) problem.

The key to an efficient implementation of the NP method was to reformulate the PMFRS such that a high quality partitioning could be obtained.

This illustrated the flexibility of the NP method in terms of how problems can be formulated. At the same time, in order to generate high-quality sample solutions, we developed a new sampling algorithm that biases the sampling towards good schedules, and a simple resource allocation improvemen heuristic.

# 7

# Feature Selection

In this chapter we present an implementation of the NP method for selecting which variables should be used by a learning algorithm. This implementation will illustrate several important aspects of the NP method. In particular, we quantify the value of developing a good method of partitioning, or what we call an intelligent partitioning method. We also show how the method can deal effectively with complex objective functions that have no closed form, and how it remains effective even if significant noise is present in the evaluation of the objective function. Finally, we explore what can be inferred from the amount of backtracking done by the algorithm, and how that information can be used to design an adaptive NP method. However, before considering these implementation issues, it is necessary to provide some background to the problem.

Many databases are massive and contain a wealth of important data that traditional business practices fall short in transforming into relevant knowledge. This has led to an increased industry and academic interest in knowledge discovery in databases, an emerging field of growing importance. The process of discovering useful information in large databases consists of numerous steps, which may include integration of data from numerous databases, manipulation of the data to account for missing and incorrect data, and induction of a model with a learning algorithm. The model is then used to identify and implement actions to take within the enterprizes. Traditionally, data mining draws heavily on both statistics and artificial intelligence, but numerous problems in data mining and knowledge discovery can also be formulated as optimization problems (Basu 1998, Bradley et al. 1999, Ólafsson et al., 2006).

All data mining starts with a set of data called the *training set*, which consists of instances describing the observed values of certain variables referred to as *features*. These instances are then used to learn a given target concept and depending upon the nature of this concept, different learning algorithms are applied. One of the most common is classification, where a learning algorithm is used to induce a model that classifies any new instances into one of two or more given categories. The primary objective is usually for the classification to be as accurate as possible, but accurate models are not necessarily useful or interesting and other measures such as simplicity and novelty are also important. In addition to classification, other common concepts to be learned include association rules, numerical prediction models, and natural clusters of the instances. Here we focus on data mining for classification where the data are nominal; that is, each feature can take only finitely many values. If the data are not nominal, we assume that one of the standard discretization techniques is applied as a preprocessing step (e.g. Fayyad and Irani 1993).

Apart from inductive learning, an important problem in knowledge discovery is analyzing the relevance of the features, usually called *feature* or *attribute subset* selection. This feature selection problem was introduced in Chapter 1 and here we explore its solution further. Feature selection involves a process for determining which features are relevant in that they predict or explain the data, and conversely, which features are redundant or provide little information (Liu and Motoda 1998). Such feature selection is commonly used as a preliminary step preceding a learning algorithm and has numerous benefits. By eliminating many of the features it becomes easier to train other learning methods, that is, computational time is reduced. Also, the resulting model may be simpler, which often makes it easier to interpret and thus more useful in practice. It is also often the case that simple models are found to generalize better when applied for prediction (Liu and Motoda 1998). Thus, a model employing fewer features is likely to score higher on many interest measures and may even score higher in accuracy. Finally, discovering which features should be kept, that is, identifying features that are relevant to the decision making, often provides valuable structural information and is therefore important in its own right.

The literature on feature selection is extensive within the machine-learning and knowledge-discovery communities. Some of the methods applied to this problem in the past include genetic algorithms (Yang and Honavar 1998), various sequential search algorithms (see, e.g., Aha and Bankert 1996, Caruna and Freitag 1994), correlation-based algorithms (Hall 2000), evolutionary search (Kim et al. 2000), rough sets theory (Modrzejewski 1993), randomized search (Skalak 1994) and branch-and-bound (Naranda and Fukunaga 1977).

Feature-selection methods are typically classified as either *filtering methods*, which produce a ranking of all features before the learning algorithm is applied, or *wrapper methods*, which use the learning algorithm to evaluate subsets of features. As a general rule, filtering methods are faster, whereas wrapper methods usually produce subsets that result in more accurate models. Another way to classify the various methods is according to whether they evaluate one feature at a time and either include or eliminate this feature, or whether an entire subset of features is evaluated together. We note that wrapper methods always fall into the latter category.

The feature-selection problem is generally difficult to solve. The number of possible feature subsets is $2^n$, where $n$ is the number of features, and evaluating every possible subset is therefore prohibitively expensive unless $n$ is very small. Furthermore, in general there is no structure present that allows for an efficient search through this large space, and a heuristic approach that sacrifices optimality for efficiency is typically applied in practice. Thus, most existing methods do not guarantee that the set of selected features is optimal in any sense. A notable exception is recent work that applies mathematical programming to feature selection (Bradley et al. 1998). Nonetheless, many of the methods mentioned above have proven themselves valuable in practice, but not being able to make rigorous statements about the set of selected features without resorting to computationally expensive or otherwise restrictive methods is an apparent shortcoming of the current state of the art.

## 7.1 NP Method for Feature Selection

The feature selection problem is a large-scale discrete optimization problem and it is clear that the NP method is applicable. The primary difficulty in solving this problem is that there is no simple method for defining an objective function, and the quality of the solution (feature subset) is often measured by its eventual performance when used with a learning algorithm. Thus, the objective function becomes the estimated performance of a complex learning algorithm. Since the NP method can deal effectively with such complex objective functions, it becomes an attractive method for solving this problem. As for other problems, the key to an efficient implementation of the NP method for feature selection is to devise intelligent partitioning (see Section 2.2) and good methods of randomly generating high quality feasible solutions (see Section 2.3). These two issues are addressed in the next two subsections.

### 7.1.1 Intelligent Partitioning

The feature selection problem was used as an example in Section 2.2.3, where we showed how to develop an intelligent partitioning method for this problem. We recall that we defined the decision variables to determine whether a feature is included in the set of selected feature, that is,

$$x_i = \begin{cases} 1 \text{ if the } i\text{th feature is included,} \\ 0 \text{ otherwise.} \end{cases}$$

Thus, given a current set $\mathcal{A}(k)$ of potential feature subsets, partition the set into two disjoint subsets

$$\mathcal{A}_1(k) = \{A \in \mathcal{A}(k) : a \in A\}, \tag{7.1}$$
$$\mathcal{A}_2(k) = \{A \in \mathcal{A}(k) : a \notin A\}. \tag{7.2}$$

Hence, a partition is defined by a sequence of features $a_1, a_2, \ldots, a_n$, which determines the order in which the features are either included or excluded (see Figure 2.5).

Also recall that for a good partitioning, it is advantageous to start by partitioning on the most important features first, and the information gain $Gain(T, a)$ of a feature $a$ relative to a training data set $T$, as defined by equation (2.11), can be used to measure the importance of features. Intuitively, this is the expected reduction in entropy $E(a)$, as defined by equation (2.9), that would occur if we knew the value of feature $a$. Note that the feature with the highest information gain has the lowest entropy value.

Thus, we obtain the entropy partition where the maximum information gain, or equivalently the minimum entropy, determines the order in which the features are used:

$$a_1 = \arg \min_{a \in A^{(ALL)}} E(a),$$

$$a_2 = \arg \min_{a \in A^{(ALL)} \setminus \{a_1\}} E(a),$$

$$\vdots$$

$$a_n = \arg \min_{a \in A^{(ALL)} \setminus \{a_1, \ldots, a_{n-1}\}} E(a),$$

where $A^{(ALL)}$ denotes the set of all features.

### 7.1.2 Generating Feasible Solutions

In addition to an intelligent partitioning that imposes good structure on the search space, the other critical component of an efficient and effective NP implementation is being able to effectively generate high-quality feasible solutions from each region (see Section 2.3). Just as the information gain (entropy) measure can be used to impose structure on the feasible region, it can also be used to bias the sampling distribution when generating feasible solutions.

Generating feasible samples can as before either be done generically using uniform sampling or by incorporating an structure that assigns different probability to different feature subsets. The latter may be expected to improve the quality of the samples and hence increase the probability that the correct region is identified in each iteration. This is our focus here. The aim isto select good feature subsets with higher probability and thus more quickly obtain a good estimate of the quality of each region. The idea of information gain can again be used. As features with high information gain are believed to be more useful, it is intuitively appealing to select those features with higher probability than features with low information gain. We thus propose the following approach for determining whether a feature should be included in a sample. Assume that we are at the $k$th iteration and, as before, let $a_1, a_2, \ldots, a_n$ denote the selected sequence of features. We let $d(k)$ denote the position or depth of $\mathcal{A}(k)$ in the partition tree of the current most promising region. Recall that this implies that the first $d(k)$ features have been fixed as either included or excluded in the current set of features. We then sample according to the following probabilities:

$$Prob[\text{Select feature } a_i] = \frac{Gain(T, a_i)}{K \max\limits_{h \in \{d(k)+1,\ldots,n\}} Gain(T, a_h)}, \qquad (7.3)$$

for $i = d(k) + 1, d(k) + 2, \ldots, n$. Here $Gain(T, a)$ is the information gain of each feature calculated according to (2.11), and $K > 1$ is a scaling constant. Note that all or none of the features can be included in the sample, and the higher the information gain, the more likely it is that a feature will be included in a sample feature set. Furthermore, by selecting $K > 1$ there is a positive probability of selecting feature subsets that do not include the feature with the highest information gain, and the expected number of features included is inversely proportional to the value of $K$.

Finally, once sample feature subsets have been obtained, these subsets must be evaluated. As discussed in the introduction, how this is done defines whether the algorithm is a filter or a wrapper approach. An important property of the NP framework is that it can be implemented according to either approach, and we thus consider two alternatives:

- **(NP wrapper)** We can use the learning algorithm itself to measure the performance of a set, i.e., the set-performance function becomes

$$f(A) = \text{Accuracy}(A), \qquad (7.4)$$

where the accuracy depends on which learning algorithm is applied. This is the wrapper approach and we refer to the NP algorithm that uses (7.4) as the *NP wrapper*.
- **(NP filter)** We can also use the correlation among features to measure the performance of each feature set. The basic idea here is that good feature sets should correlate highly with the class feature, but have low correlation with each other. We can thus use the following performance function (Hall 2000):

$$f(A) = \frac{k \bar{\rho}_{ca}}{\sqrt{k + k(k-1)\bar{\rho}_{aa}}}, \qquad (7.5)$$

where $k$ is the number of features in the set $A$, $\bar{\rho}_{ca}$ is the average correlation between the features in this set and the classification feature, and $\bar{\rho}_{aa}$ is the average correlation between features in the set $A$. We note that this is a filter approach and hence we refer to it as the *NP filter*.

We note that any learning algorithm can be used with a NP wrapper, and methods other than the correlation measure (7.5) that similarly evaluate feature subsets can be used for different variants of the NP filter.

## 7.2 NP-Wrapper and NP-Filter Algorithm

### 7.2.1 NP Filter Algorithm

With all of the components of NP for feature selection in place, we can state the proposed feature-selection algorithms completely. The following algorithm can be used to implement the filter approach. Note that it uses a fixed number $n_0$ of samples to evaluate each region, starts with the set $\mathcal{A}$ of all possible feature subsets as the most promising region, and terminates when the depth of the most promising region has reached a maximum, i.e., it is a singleton. We also let $A^*$ be the best feature subset found and $f^*$ be the corresponding performance value, which is calculated according to (7.5).

**Algorithm** *NP Filter*

Step 0.   Select the constant $K > 1$ for scaling the sampling distribution, and select $n_0$, the number of sample points. Evaluate the entropy value of each feature and let $a_1, a_2, ..., a_n$ be the corresponding order of features:

$$a_1 = \arg \min_{a \in A^{(ALL)}} E(a),$$

$$a_2 = \arg \min_{a \in A^{(ALL)} \backslash \{a_1\}} E(a),$$

$$\vdots$$

$$a_n = \arg \min_{a \in A^{(ALL)} \backslash \{a_1, ..., a_{n-1}\}} E(a).$$

Set $\mathcal{A}(0) = \mathcal{A}$, $k = 0$, and $d(0) = 0$. Let $A^* = \{\}$ and set $f^* = \infty$.

Step 1.   Partition $\mathcal{A}(k)$ into two subregions and aggregate what remains into one complimentary region:

$$\mathcal{A}_1(k) = \{A \in \mathcal{A}(k) : a_{d(k)} \in A\},$$
$$\mathcal{A}_2(k) = \{A \in \mathcal{A}(k) : a_{d(k)} \notin A\},$$
$$\mathcal{A}_3(k) = \mathcal{A} \backslash \mathcal{A}(k).$$

Step 2.   From each of the three regions, independently obtain $n_0$ sample sets $A_1^j, A_2^j, ..., A_{n_0}^j$, $j = 1, 2, 3$, according the the distribution

$$Prob[\text{Select feature } a_i] = \frac{Gain(T, a_i)}{K \max\limits_{h \in \{d(k)+1, ..., n\}} Gain(T, a_h)},$$

for $i = d(k) + 1, d(k) + 2, \ldots, n$. Here $Gain(T, a)$ is calculated according to (2.11).

Step 3.    Obtain the best sample set from each region

$$A_{best}^j = \arg \min_{l=1,2,\ldots,n_0} f\left(A_l^j\right),$$

and $f(\cdot)$ is defined according to (7.5).

Step 4.    Select the next most-promising region based on the sample results
   (a)    If $f(A_{best}^1) < \min\{f(A_{best}^2), f(A_{best}^3)\}$, let $\mathcal{A}(k+1) = \mathcal{A}_1(k)$
          and $d(k+1) = d(k) + 1$.
          If $f(A_{best}^1) < f^*$, let $f^* = f(A_{best}^1)$ and $A^* = A_{best}^1$.
   (b)    If $f(A_{best}^2) < \min\{f(A_{best}^1), f(A_{best}^3)\}$, let $\mathcal{A}(k+1) = \mathcal{A}_2(k)$
          and $d(k+1) = d(k) + 1$.
          If $f(A_{best}^2) < f^*$, let $f^* = f(A_{best}^2)$ and $A^* = A_{best}^2$.
   (c)    Otherwise, let $\mathcal{A}(k+1) = s(\mathcal{A}_k)$ where $s(\mathcal{A}(k))$ is the super-
          region of $\mathcal{A}(k)$, and $d(k+1) = d(k) - 1$.
          If $f(A_{best}^3) < f^*$, let $f^* = f(A_{best}^3)$ and $A^* = A_{best}^3$.
Step 5.    If $d(k+1) = n$, stop and return $A^*$ as the best subset. Otherwise,
           let $k = k + 1$ and go back to Step 1.

Note that the algorithm requires that the user to select a value of the scaling constant $K$ for the probability distribution. No absolute rule can be given for the selection of $K$ but normally one would want to select it large enough so that there is a significant probability that the highest-information-gain feature is not selected in every random set, and that there is a sufficiently high probability of selecting other features. For example, if we let $K = 1.25$ then the probability of selecting the feature with the highest information gain is 80% for each subset generated.

Except for slight modifications to Step 3, the NP wrapper is implemented in an identical fashion. We therefore do not present that implementation in detail, but rather illustrate this algorithm via an example.

### 7.2.2 NP Wrapper Example

To illustrate the mechanism of the new algorithm we apply the NP wrapper using naive Bayes as a learning algorithm to the simple weather-classification problem illustrated in Table 7.1. Here there are four features (*Outlook, Temperature, Humidity,* and *Windy*) that can be used to predict a class feature that can take two values: play or no play. To determine the order in which features are selected for partitioning, we calculate the entropy of each feature according to (2.9):

$$E(Outlook) = 0.693,$$
$$E(Temperature) = 0.911,$$
$$E(Humidity) = 0.788,$$
$$E(Windy) = 0.892.$$

**Table 7.1.** Data for simple weather example.

| Outlook | Temperature | Humidity | Windy | Class |
|---------|-------------|----------|-------|-------|
| sunny | hot | high | false | no play |
| sunny | hot | high | true | no play |
| overcast | hot | high | false | play |
| rain | mild | high | false | play |
| rain | cool | normal | false | play |
| rain | cool | normal | true | no play |
| overcast | cool | normal | true | play |
| sunny | mild | high | false | no play |
| sunny | cool | normal | false | play |
| rain | mild | normal | false | play |
| sunny | mild | normal | true | play |
| overcast | mild | high | true | play |
| overcast | hot | normal | false | play |
| rain | mild | high | true | no play |

The expected information of the training set $T$ is $I(T) = 0.94$, so *Outlook* has the highest information gain $Gain(T, Outlook) = 0.247$, followed by *Humidity* with $Gain(T, Humidity) = 0.152$ and *Windy* with $Gain(T, Windy) = 0.048$, and finally *Temperature* has the smallest information gain

$$Gain(T, Temperature) = 0.029.$$

The resulting partitioning tree is shown in Figure 7.1, which also shows the final feature set for each maximum-depth region, and the corresponding accuracy value when naive Bayes is used as a learning algorithm. As this problem is quite small we have chosen the maximum depth equal to the total number of features. The accuracy values in Figure 7.1, as well as those used by the NP wrapper, are calculated using ten-fold cross-validation.

First note that intelligent partitioning indeed imposes a useful structure on the space of feature subsets, which can be exploited by the NP search. As discussed in Section 2.2 we can measure this in two ways: using the variability of the accuracy or the percentage overlap between the set containing the global optimum and other sets. In particular, note that feature subsets with similar accuracy tend to be grouped together and the sample estimates of the best accuracy in each region will therefore have low variability. An extreme case is the set defined by all feature subsets containing *Outlook* but not containing *Humidity*, where every feature subset has the same accuracy and hence the variability is zero. Similarly, the overlap is small and, in the extreme case, the set containing *Outlook* and *Humidity*, which includes the optimum, has no overlap with other subsets. Due to this imposed structure it can be expected that the random search quickly moves toward the optimal feature subset $\{Outlook, Humidity\}$.

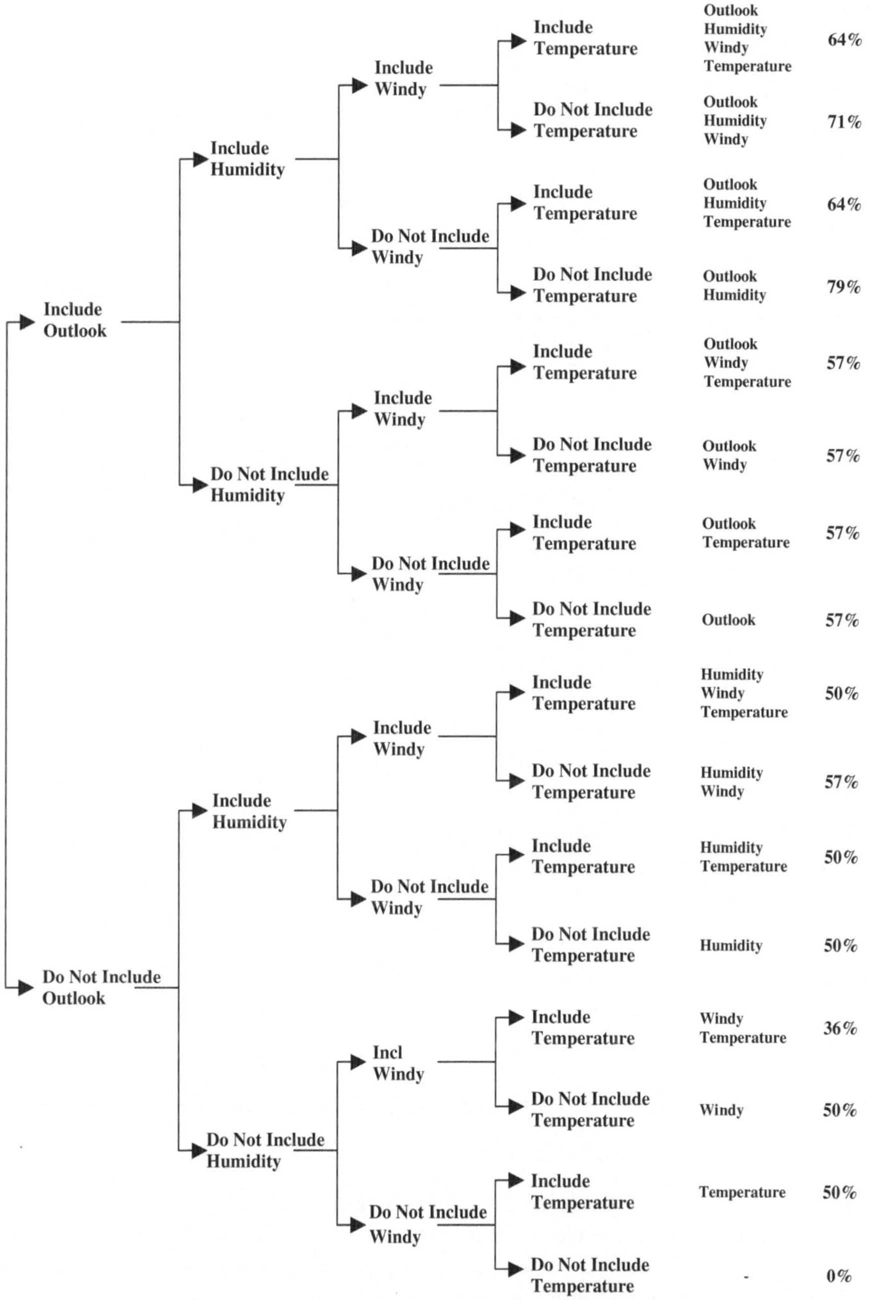

| | | | Include<br>Temperature | Outlook<br>Humidity<br>Windy<br>Temperature | 64% |
| Include<br>Windy | | | Do Not Include<br>Temperature | Outlook<br>Humidity<br>Windy | 71% |
| Include<br>Humidity | | | Include<br>Temperature | Outlook<br>Humidity<br>Temperature | 64% |
| Do Not Include<br>Windy | | | Do Not Include<br>Temperature | Outlook<br>Humidity | 79% |

**Fig. 7.1.** Partitioning tree for weather example.

We now illustrate a few iterations of the algorithm. We initialize the algorithm by setting $\mathcal{A}(0) = \mathcal{A}$, and then partition and sample as follows: The most promising region $\mathcal{A}(0)$ is partitioned into two subsets depending on whether we include or exclude the feature with the highest information gain, namely *Outlook*:

$$\mathcal{A}_1(0) = \{A \in \mathcal{A}(0) : Outlook \in A\},$$
$$\mathcal{A}_2(0) = \{A \in \mathcal{A}(0) : Outlook \notin A\}.$$

Next we obtain samples from each of those regions, according to the distribution (7.3) with $K = 1.2$, which takes the information gain into account. For example, the probabilities that each of the three remaining features is included in a sample from $\mathcal{A}_1(0)$ are given as

$$Prob[\text{Sample includes } Temperature] = \frac{0.029}{1.2 \cdot 0.152} = 0.16$$
$$Prob[\text{Sample includes } Humidity] = \frac{0.152}{1.2 \cdot 0.152} = 0.83$$
$$Prob[\text{Sample includes } Windy] = \frac{0.048}{1.2 \cdot 0.152} = 0.26.$$

Thus, for each sample obtained, it has the following distribution

$$Prob[\text{Select } \{Outlook, Temperature\}] = 0.16 \cdot (1 - 0.83) \cdot (1 - 0.26)$$
$$= 0.02$$
$$Prob[\text{Select } \{Outlook, Humidity\}] = (1 - 0.16) \cdot 0.83 \cdot (1 - 0.26)$$
$$= 0.52$$
$$Prob[\text{Select } \{Outlook, Temperature, Humidity\}] = 0.16 \cdot 0.83 \cdot (1 - 0.26)$$
$$= 0.10$$
$$\vdots$$

Thus, each of the eight feature subsets has a positive probability of being randomly selected, but the probability depends on the perceived information gain of the features in the set.

Say that we select one sample from each region, $\{Outlook, Windy, Temperature\}$ from $\mathcal{A}_1(0)$ and $\{Humidity, Windy\}$ from $\mathcal{A}_2(0)$. Since

$$f(\{Outlook, Windy, Temperature\}) = 57 = f(\{Humidity, Windy\}),$$

a tie must be broken. Given the goals of feature selection, we adopt the rule of breaking ties by favoring the smaller set, i.e., $\{Humidity, Windy\}$, so in the next iteration $\mathcal{A}(1) = \mathcal{A}_2(0)$ and this new most-promising region is partitioned into two subregions, and what remains is aggregated into one set:

$$\mathcal{A}_1(1) = \{A \in \mathcal{A} : Outlook \notin A, Humidity \in A\},$$
$$\mathcal{A}_2(1) = \{A \in \mathcal{A} : Outlook \notin A, Humidity \notin A\},$$
$$\mathcal{A}_3(1) = \mathcal{A} \setminus \mathcal{A}(1).$$

A quick glance at Figure 7.1 reveals that this move takes the search away from the optimal solution, but by maintaining the complimentary region $\mathcal{A}_3(1)$ the algorithm is able to recover. Thus, we· obtain one sample from each region, say, $\{Humidity, Windy\}$ from $\mathcal{A}_1(1)$, $\{Windy\}$ from $\mathcal{A}_2(1)$, and $\{Outlook, Humidity\}$ from $\mathcal{A}_3(1)$. As the sample from $\mathcal{A}_3(1)$ has the best accuracy, the algorithm backtracks and sets $\mathcal{A}(2) = \mathcal{A}$.

We are now back where we started and this time we are likely to select different samples, say $\{Outlook, Humidity, Temperature\}$ from $\mathcal{A}_1(2)$ and $\{Humidity, Windy\}$ from $\mathcal{A}_2(2)$. This time around,

$$f(\{Outlook, Humidity, Temperature\}) = 64 > 57 = f(\{Humidity, Windy\}),$$

and the first subset is selected as the most-promising region, i.e., $\mathcal{A}(3) = \mathcal{A}_1(2)$. It is partitioned into two subregions, and what remains is aggregated as before:

$$\mathcal{A}_1(3) = \{A \in \mathcal{A} : Outlook \in A, Humidity \in A\},$$
$$\mathcal{A}_2(3) = \{A \in \mathcal{A} : Outlook \in A, Humidity \notin A\},$$
$$\mathcal{A}_3(3) = \mathcal{A} \setminus \mathcal{A}(3).$$

Now note that, regardless of which samples are selected from these three regions, the sample from $\mathcal{A}_1(3)$ will have the highest accuracy, so $\mathcal{A}(4) = \mathcal{A}_1(3)$. The new most-promising region is partitioned into two subregions and what remains is aggregated into one set:

$$\mathcal{A}_1(4) = \{\{Outlook, Humidity, Windy, Temperature\},$$
$$\{Outlook, Humidity, Windy\}\},$$
$$\mathcal{A}_2(4) = \{\{Outlook, Humidity, Temperature\}, \{Outlook, Humidity\}\},$$
$$\mathcal{A}_3(4) = \mathcal{A} \setminus \mathcal{A}(4).$$

Depending on the sampling, either $\mathcal{A}_1(4)$ or $\mathcal{A}_2(4)$ may be selected, but since there is no overlap between $\mathcal{A}_3(4)$ and the "good" region $\mathcal{A}_2(4)$, backtracking will not be warranted by the sampling.

From these first four iterations it is clear that the sequence of most promising regions moves towards the optimum with high probability and has the potential to recover from wrong moves via backtracking. As the algorithm progresses, the sampling, and thus the computational effort, is concentrated where good feature subsets are likely to be found. Once the maximum depth is reached, that is, the current most promising region is a singleton, the algorithm stops. Although simple, this example thus illustrates many key aspects of the NP method, including how it converges, how the computational effort is focused in most promising regions, the value of backtracking, and the paramount importance of an intelligent partitioning strategy.

## 7.3 Numerical Comparison with Other Methods

In this section we present numerical results for tests of the NP wrapper and NP filter when used to precede two classification algorithms, namely the naive Bayes algorithm and the C4.5 decision tree induction algorithm. The code was written in Java using the *Weka* machine learning software (Witten et al. 1999) for implementation of the learning algorithms themselves. We use five data sets from the *UCI repository of machine learning databases* (Blake and Merz 1998). The characteristics of these sets are shown in Table 7.2, from which we note that the sizes range from 148 to 3196 instances and from nine to 69 features. As both the NP filter and NP wrapper are randomized algorithms, we ran five replications for each experiment and report both the average and the standard error. All accuracy estimates are obtained using ten-fold cross-validation, and the averages reported therefore average five cross-validation estimates.

### 7.3.1 Value of Feature Selection

Our first set of experiments addresses the effectiveness of feature selection using the NP filter and NP wrapper for the selected data sets. As noted before, both naive Bayes and C4.5 are used to induce classification models with the selected features. We measure the effectiveness along two dimensions. First, we consider the accuracy of the models induced after feature selection compared to the corresponding models without feature selection, and second, we consider how many features are eliminated, i.e., how much smaller the models become when feature selection is employed.

The results for the naive Bayes classification method are shown in Table 7.3 for no feature selection (NFS), the NP filter (NPF), and the NP wrapper (NPW). Improvements that a $t$-test with four degrees of freedom finds to be statistically significant at the 90% level are marked with an asterisk. Looking first at the accuracy, we note that it actually improves or is no worse when we use feature selection, and the models where classification is preceded by an NP wrapper have the highest accuracy. All but one of the NPW models show a statistically significant improvement. Table 7.3 demonstrates the reduction in the number of features. For example, when the NP filter is used, the 69

**Table 7.2.** Characteristics of the tested data sets.

| Data Set | Instances | Features |
|----------|-----------|----------|
| lymph | 148 | 18 |
| vote | 435 | 16 |
| audiology | 226 | 69 |
| cancer | 286 | 9 |
| kr-vs-kp | 3196 | 36 |

**Table 7.3.** Accuracy of Naive Bayes with and without feature selection.

| | NFS | | NPF | | NPW | |
|---|---|---|---|---|---|---|
| Data Set | Accuracy | Size | Accuracy | Size | Accuracy | Size |
| lymph | 85.1 | 18 | 85.4±1.0 | 10.6±2.1 | 86.2±0.8 | 9.2±0.8 |
| vote | 90.1 | 16 | 93.2±0.7* | 6.8±1.1 | 95.8±0.4* | 3.0±1.4 |
| audiology | 71.2 | 69 | 71.2±1.5 | 27.4±3.2 | 75.0±2.3* | 23.0±3.9 |
| breast-cancer | 73.4 | 9 | 73.8±0.4 | 5.8±0.8 | 75.7±0.2* | 3.6±0.9 |
| kr-vs-kp | 88.0 | 36 | 90.8±2.1 | 11.6±1.5 | 94.4±0.3* | 14.2±3.8 |

\* *Statistically significant improvement over NFS.*

**Table 7.4.** Accuracy of C4.5 with and without feature selection.

| | NFS | | NPF | | NPW | |
|---|---|---|---|---|---|---|
| Data Set | Accuracy | Size | Accuracy | Size | Accuracy | Size |
| lymph | 78.4 | 18 | 78.1±1.4 | 9.6±1.8 | 82.2±0.4* | 7.6±1.7 |
| vote | 96.5 | 16 | 95.7±0.2 | 6.0±2.0 | 96.6±0.6 | 3.8±1.6 |
| audiology | 77.4 | 69 | 75.0±1.6 | 25.0±4.3 | 79.9±1.1* | 14.0±2.7 |
| cancer | 75.5 | 9 | 73.7±0.2 | 5.2±0.5 | 76.2±0.6 | 2.4±0.9 |
| kr-vs-kp | 99.1 | 36 | 94.0±1.1 | 11.6±1.3 | 96.5±0.9 | 17.0±2.7 |

\* *Statistically significant improvement over NFS.*

features of the "audiology" data set are reduced to an average of 27.4 features, and when the NP wrapper is used they are reduced to an average of 23.0 features. This is a significant simplification of the models. We note that the NP wrapper performs better on both the accuracy and simplicity measures.

We repeat the same experiments for the C4.5 decision-tree-induction algorithm, with the results reported in Table 7.4 according to the same format as before. Here the accuracy of the model is actually degraded somewhat when the NP filter is used, but using the NP wrapper still results in higher-accuracy models than when using all of the features for all but one of the data sets, although this is statistically significant at the 90% level for only two of the five models. Thus, there is relatively little accuracy gain from using feature selection with C4.5. This is not surprising as this decision-tree-induction method already employs a sophisticated approach to both selecting the order of features in the tree and to pruning the tree afterwards. The reduction in the number of features, however, is even larger than before.

### 7.3.2 Comparison with Simple Entropy Filter

The last subsection demonstrated the value of using the NP filter and NP wrapper for feature selection as this results in smaller but often higher-accuracy models. However, the question arises as to whether this performance is primarily due to properties of the NP method or because it incorporates an

**Table 7.5.** Accuracy of Naive Bayes with early termination of feature selection.

| Data Set | Depth | NPF Accuracy | Size | NPW Accuracy | Size | EF Accuracy | Size |
|---|---|---|---|---|---|---|---|
| lymph | | 84.7±1.6* | 11.6±1.1 | 86.9±1.9* | 9.6±0.6 | 81.8 | 11 |
| vote | | 93.0±0.5* | 6.8±0.8 | 95.0±0.6* | 4.8±1.3 | 89.9 | 10 |
| audiology | Max | 69.7±0.7 | 28.6±3.3 | 73.6±1.6 | 30.0±3.0 | 75.2 | 43 |
| cancer | | 73.7±0.3 | 5.4±0.6 | 75.5±0.5* | 2.8±0.8 | 74.1 | 6 |
| kr-vs-kp | | 90.7±1.2* | 11.6±0.9 | 94.1±0.4* | 13.0±1.6 | 88.1 | 23 |
| lymph | | 83.9±0.7* | 10.6±0.9 | 86.4±0.3* | 11.2±1.1 | 81.8 | 7 |
| vote | | 93.2±0.4* | 8.2±1.1 | 94.7±0.8* | 3.8±0.8 | 92.4 | 6 |
| audiology | Avg | 71.0±2.7 | 24.8±3.5 | 74.8±2.8 | 26.6±4.3 | 73.5 | 26 |
| cancer | | 73.6±0.3* | 4.8±0.5 | 75.4±0.8* | 3.0±1.7 | 72.7 | 3 |
| kr-vs-kp | | 90.8±2.8 | 10.0±2.1 | 93.2±0.7* | 12.4±1.5 | 89.9 | 14 |
| lymph | | 85.7±0.7* | 11.8±1.6 | 86.0±1.8* | 9.4±0.9 | 80.4 | 2 |
| vote | | 99.0±1.4* | 5.4±2.5 | 94.3±0.5 | 5.8±0.5 | 95.6 | 2 |
| audiology | Min | 70.5±3.1 | 11.6±1.8 | 73.5±2.5* | 14.0±4.3 | 67.7 | 9 |
| cancer | | 73.7±0.4* | 4.6±0.6 | 74.3±0.9* | 4.2±1.6 | 72.0 | 1 |
| kr-vs-kp | | 90.7±1.3* | 7.6±0.9 | 94.2±0.1* | 7.4±1.7 | 86.7 | 5 |

*\* Statistically significant improvement over EF.*

information gain ranking of features that is known to perform quite well in practice.

To evaluate the contribution of the NP method versus that of simply using the information-gain ranking we compare the performance of the NP filter and NP wrapper with a filter that we refer to as the *entropy filter* (EF). This filter simply selects the features with the highest information gain to be included in the model. Since the number of features used by the solutions found by the NP filter and NP wrapper is not fixed, a comparison of models with the same number of features is not possible. However, for a fairer comparison we change the stopping criterion of the NP algorithms so that we terminate when a certain depth is reached, i.e., after a given number of features has been considered for inclusion in the set. The same number of features is then used by the EF. Given the data sets have a varying number of features but that a common testing procedure is desired, we consider using approximately 60%, 40%, and 15% of the features in three different experiments. The results for naive Bayes are shown in Table 7.5. The first set of results for each data set uses 60% of features (Max), the second set 40% (Avg), and the third 15% (Min). From these results we see that the simple EF actually performs quite well, but on average both the NP filter and NP wrapper perform significantly better with respect to accuracy. The average accuracy of the NP filter is better for all problems except the audiology test set. The average accuracy of the NP wrapper is better for all problems, and in 13 out of 15 experiments the average accuracy of the NP wrapper is better than for both of the other methods.

**Table 7.6.** Accuracy of C4.5 with early termination of feature selection.

|  |  | NPF | | NPW | | EF | |
|---|---|---|---|---|---|---|---|
|  | Depth | Accuracy | Size | Accuracy | Size | Accuracy | Size |
| lymph |  | 78.1±0.8* | 12.0±0.7 | 82.2±1.0* | 8.4±2.3 | 76.4 | 9 |
| vote |  | 95.5±0.2 | 6.4±1.8 | 96.4±0.7 | 6.0±2.6 | 95.6 | 8 |
| audiology | Max | 76.4±0.9 | 28.6±3.5 | 79.6±1.7 | 27.4±4.2 | 77.9 | 36 |
| cancer |  | 73.4±0.0 | 6.0±0.0 | 75.9±0.2* | 4.2±0.5 | 73.8 | 5 |
| kr-vs-kp |  | 91.6±3.6 | 10.0±0.7 | 96.7±0.8 | 17.2±2.8 | 97.1 | 19 |
| lymph |  | 79.1±1.4 | 10.6±1.1 | 81.5±0.6* | 9.0±0.7 | 78.4 | 6 |
| vote |  | 96.6±0.2* | 8.2±1.6 | 96.3±0.5 | 7.2±1.9 | 95.6 | 5 |
| audiology | Avg | 74.6±2.9 | 27.2±2.2 | 79.5±1.6 | 30.4±2.5 | 77.9 | 22 |
| cancer |  | 73.8±0.0* | 5.0±0.0 | 75.9±0.0* | 2.6±0.9 | 71.7 | 3 |
| kr-vs-kp |  | 91.9±4.2 | 11.8±2.1 | 97.1±0.3* | 1.8±0.8 | 96.5 | 12 |
| lymph |  | 78.7±0.6* | 9.0±0.7 | 81.1±0.5* | 9.8±1.9 | 73.7 | 2 |
| vote |  | 95.5±0.3 | 6.2±1.9 | 96.3±0.5 | 7.2±1.9 | 95.6 | 2 |
| audiology | Min | 73.2±2.1* | 13.0±3.1 | 77.2±1.5* | 16.4±2.4 | 69.9 | 8 |
| cancer |  | 72.9±1.2* | 4.6±0.6 | 74.1±1.1* | 3.2±1.3 | 69.6 | 1 |
| kr-vs-kp |  | 87.2±6.2 | 6.6±0.9 | 96.2±0.9* | 13.2±3.1 | 90.4 | 4 |

\* *Statistically significant improvement over EF.*

The same results for the C4.5 decision tree algorithm are shown in Table 7.6. The results here are similar, except that the NP filter performs relatively worse, with an average improvement for only three of the five problems, namely for the "lymph," "vote," and "cancer" data sets. Again, the NP wrapper has the best performance for 13 out of 15 experiments.

The improved accuracy obtained by using the NP algorithms, and especially the NP wrapper, does of course come at a price, which is increased computational time. In Tables 7.7 and 7.8 we report the amount of computation time (in milliseconds) used by each of the algorithms for all experiments reported above. For each of the data sets the first line reports the time used for the experiments reported in Tables 7.3 and 7.4 and the next three lines report the time used for the experiments reported in Tables 7.5 and 7.6.

From these results we see that using the EF takes the least amount of time, followed by using no feature selection at all. Thus, even though using the EF adds a step to the process, this is more than compensated for by the faster induction of the classification model that occurs when fewer features are employed. The NP wrapper takes by far the most amount of computation time and the NP filter falls between the NP wrapper and no feature selection.

### 7.3.3 The Importance of Intelligent Partitioning

We have seen that the high accuracy obtained by the NP filter and NP wrapper is not completely explained by the use of an information-gain ranking. Conversely, we can ask how much is due to the generic NP framework itself

**Table 7.7.** Average speed using Naive Bayes (milliseconds).

| Data Set | Depth | NFS | NPF | NPW | EF |
|---|---|---|---|---|---|
| lymph | Full | 114 | 5313 | 6113 | N/A |
| | Max | N/A | 3954 | 3833 | 120 |
| | Avg | N/A | 2557 | 2628 | 120 |
| | Min | N/A | 1190 | 853 | 112 |
| vote | Full | 164 | 11541 | 13341 | N/A |
| | Max | N/A | 8096 | 8711 | 180 |
| | Avg | N/A | 5416 | 5350 | 178 |
| | Min | N/A | 1791 | 1915 | 154 |
| audiology | Full | 370 | 124205 | 127769 | N/A |
| | Max | N/A | 88125 | 77452 | 322 |
| | Avg | N/A | 52583 | 43256 | 297 |
| | Min | N/A | 16049 | 16387 | 252 |
| cancer | Full | 119 | 2390 | 2888 | N/A |
| | Max | N/A | 1679 | 1987 | 124 |
| | Avg | N/A | 1232 | 1064 | 124 |
| | Min | N/A | 573 | 431 | 126 |
| kr-vs-kp | Full | 886 | 410682 | 515299 | N/A |
| | Max | N/A | 272552 | 315520 | 1294 |
| | Avg | N/A | 160109 | 192314 | 1298 |
| | Min | N/A | 50278 | 63697 | 1262 |

**Table 7.8.** Average speed using C4.5 (milliseconds).

| Data Set | Depth | NFS | NPF | NPW | EF |
|---|---|---|---|---|---|
| lymph | Full | 304 | 5476 | 28929 | N/A |
| | Max | N/A | 4210 | 14188 | 270 |
| | Avg | N/A | 2774 | 9514 | 276 |
| | Min | N/A | 1402 | 2912 | 240 |
| vote | Full | 392 | 11685 | 41708 | N/A |
| | Max | N/A | 8238 | 22178 | 356 |
| | Avg | N/A | 5588 | 13890 | 302 |
| | Min | N/A | 2017 | 5454 | 222 |
| audiology | Full | 1076 | 127100 | 327725 | N/A |
| | Max | N/A | 92225 | 172884 | 845 |
| | Avg | N/A | 54246 | 100969 | 709 |
| | Min | N/A | 16622 | 32048 | 488 |
| cancer | Full | 373 | 2650 | 8142 | N/A |
| | Max | N/A | 1985 | 4877 | 303 |
| | Avg | N/A | 1538 | 2940 | 284 |
| | Min | N/A | 801 | 1031 | 188 |
| kr-vs-kp | Full | 4558 | 430171 | 1836724 | N/A |
| | Max | N/A | 276450 | 614206 | 3509 |
| | Avg | N/A | 189066 | 382934 | 2794 |
| | Min | N/A | 54348 | 101643 | 1772 |

and how much to the intelligent partitioning scheme. Thus, we compare NP algorithms using the intelligent partitioning to NP algorithms using all other possible ways of partitioning . We use a complete enumeration of all partitions to find the best and worst one and compare those to intelligent partitioning. Since the number of ways in which the features can be ordered is $n!$, where $n$ is the number of features, a study of even the smallest test problem would be prohibitively time-consuming. We thus modify the data sets so that we first draw a sample of seven features and then apply the NP algorithms, repeating it five times to reduce any bias. We restrict ourselves to the "vote" and "cancer" data sets and consider only the NP wrapper with naive Bayes classification. Results for other configurations are similar.

In Table 7.9 the prediction accuracy of the models using intelligent partition, as well as the best and worst partition found using enumeration, are reported. The accuracy found using the intelligent partition is very close to optimal. For half of the problems the intelligent partitioning always results in the same accuracy as does the optimal partition, and for the other half the performance difference is not statistically significant. On the other hand, poor partitioning results in feature subsets that have significantly lower accuracy but even for the worst possible partition, the NP method is still able to obtain fairly high-quality subsets.

We also compare the computational time used by the NP wrapper if different partitioning schemes are used. These results are reported in Table 7.10 and we see again that using the intelligent partitioning results in a performance that is close to optimal, although there is greater difference than with respect to accuracy. In particular, the intelligent partitioning takes on average 1.36 and 1.75 times longer than does the best, but the worst partitioning takes on average 6.24 and 5.32 times longer than the intelligent partitioning for the the 'vote' and 'cancer' data sets, respectively. We conclude that the NP

**Table 7.9.** Accuracy evaluation of intelligent partitioning in NP Wrapper using Naive Bayes.

| Data Set | | Accuracy | | |
|---|---|---|---|---|
| | | Intelligent | Best | Worst |
| vote | 1 | 90.8±0.3 | 91.0 | 86.4 |
| | 2 | 95.9±0.0 | 95.9 | 94.3 |
| | 3 | 89.0±0.0 | 89.0 | 87.4 |
| | 4 | 90.0±0.3 | 90.1 | 85.1 |
| | 5 | 95.6±0.0 | 95.6 | 92.0 |
| cancer | 1 | 75.9±0.0 | 75.9 | 72.7 |
| | 2 | 75.7±0.2 | 75.9 | 72.7 |
| | 3 | 75.8±0.2 | 75.9 | 73.1 |
| | 4 | 72.8±0.3 | 73.1 | 70.6 |
| | 5 | 75.9±0.0 | 75.9 | 72.0 |

**Table 7.10.** Speed evaluation of intelligent partitioning in NP Wrapper using Naive Bayes.

| Data Set | | Computation Time | | |
|---|---|---|---|---|
| | | Intelligent | Slowest | Fastest |
| | 1 | 3812 | 20370 | 2184 |
| | 2 | 3515 | 19408 | 2153 |
| vote | 3 | 3371 | 35080 | 2784 |
| | 4 | 3690 | 18046 | 2774 |
| | 5 | 3433 | 17375 | 3094 |
| | 1 | 2740 | 14050 | 1382 |
| | 2 | 2624 | 16013 | 1372 |
| cancer | 3 | 2664 | 20390 | 1342 |
| | 4 | 4969 | 25226 | 1362 |
| | 5 | 2642 | 6920 | 1372 |

filter compensates fairly well for poor partitioning in terms of accuracy, but requires much computation. The intuitive explanation is that any NP algorithm can compensate for mistakes by backtracking but frequent backtracking is time-consuming and slows the search. Finally, we note that the difficulty of obtaining the optimal partitioning is in general equal to solving the problem itself. However, our results show that a very high-quality partition can be obtained efficiently with the new intelligent partitioning method.

### 7.3.4 Scalability of NP Filter

In this section we consider the scalability of the NP method for feature selection by evaluating the accuracy and computational time as functions of both the number of features and number of instances. For these tests we generated synthetic test data sets with $m \in \{50, 100, 200, 400, 800\}$ instances, and $n \in \{50, 100, 200, 400, 800\}$ features using the following approach. To create a single instance $i$, a value for the class feature $Y_i$ is generated from a uniform distribution on $[-3, 3]$, i.e., $Y_i \sim U(-3, 3)$. The value for each of the other features $X_{ij}$ is then generated via

$$X_{ij} = \rho_j Y_i + (|\rho_j| - 1) \cdot Z_j,$$

where $\rho_j$ is the amount of correlation between feature $j$ and the class feature, and $Z_j$ is drawn from a unit normal distribution, $j = 1, 2, ..., n$, $i = 1, 2, ..., m$. For each of the test problems, 10% of the features are highly correlated with $|\rho_j| \geq 0.9$, 40% have correlation $0.3 \leq |\rho_j| < 0.9$, and 50% of the features do not correlate highly with the class feature, i.e. $|\rho_j| < 0.3$. The NP filter, followed by naive Bayes classification-model induction, is run five times for each of those test sets.

First consider the scalability with respect to the number of attributes. Figure 7.2 shows accuracy as a function of the number of attributes for two

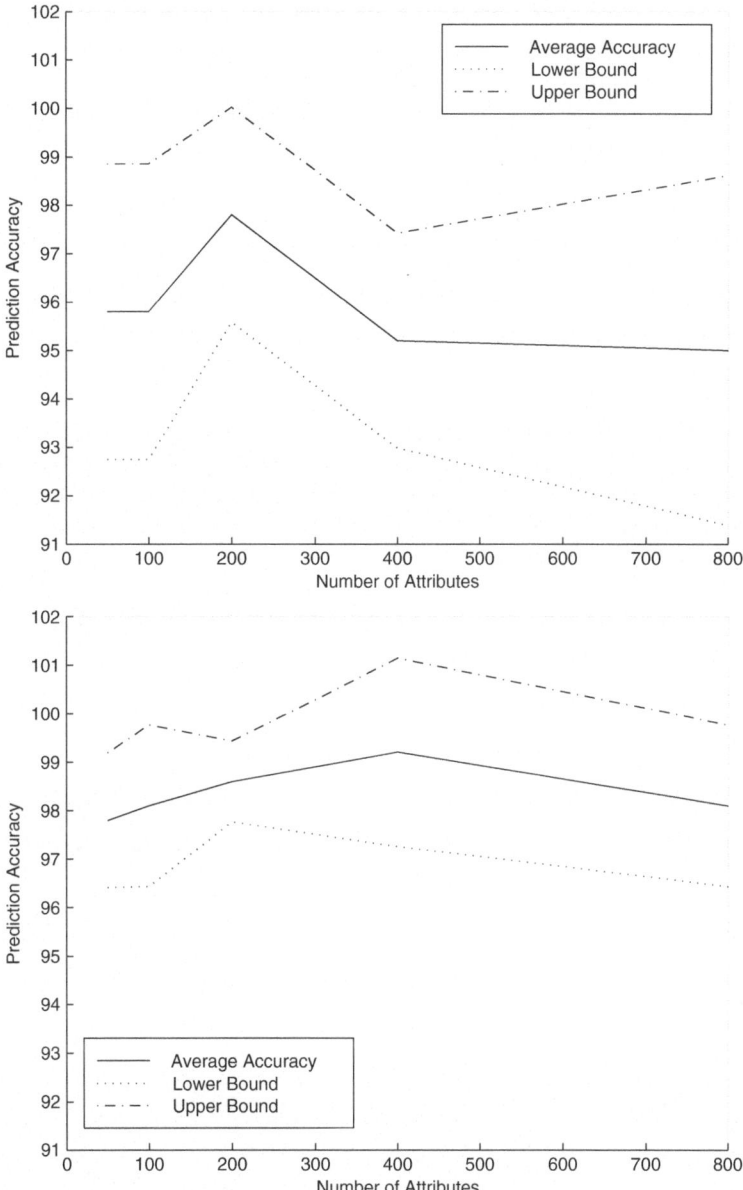

**Fig. 7.2.** Accuracy as a function of attributes for 100 instances (top) and 400 instances (bottom).

of the five instance settings (100 and 400 instances). The other settings had similar results. The figure reports both the average accuracy over the five replications (solid line) and a $t$-distribution based 95% confidence interval around this average (dashed lines). From these results we conclude that there is no significant change in the accuracy obtained as the number of attributes grows. The results for computation time as a function of attributes are reported in Figure 7.3 for three settings, 50, 200, and 800 instances. The time grows very rapidly as the number of attributes increases, and appears to demonstrate exponential growth. Thus, although quality is not lost as the problem size increases, the time it takes to achieve this quality increases quickly and the NP filter is therefore somewhat lacking in terms of scalability with respect to the number of attributes.

Finally, looking at the scalability of the NP filter as a function of the number of instances, Figure 7.4 reports the accuracy obtained as a function of the number of instances in a similar manner to what was done above. An interesting observation from this figure is that the solution quality actually improves as the problem size increases, which is not entirely surprising as more instances imply more data are available to induce a model with high accuracy. Turning now to the computation time required to achieve this accuracy, Figure 7.5 shows the time used as a function of number of instances for the 100, 200, 400, and 800 attribute problems. As opposed to the rapid growth in computation time seen when the number of attributes increases, the time here grows only linearly, thus implying that the NP filter is scalable with respect to the number of instances.

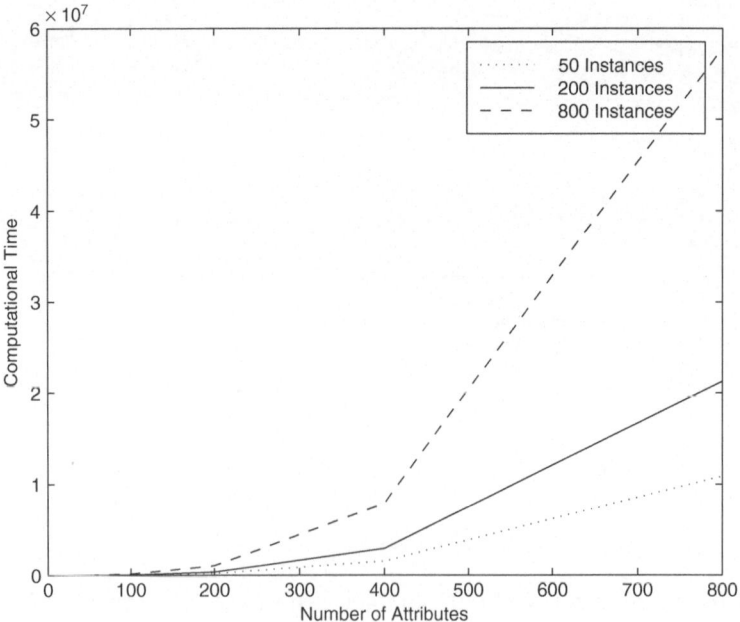

**Fig. 7.3.** Computation time as a function of attributes.

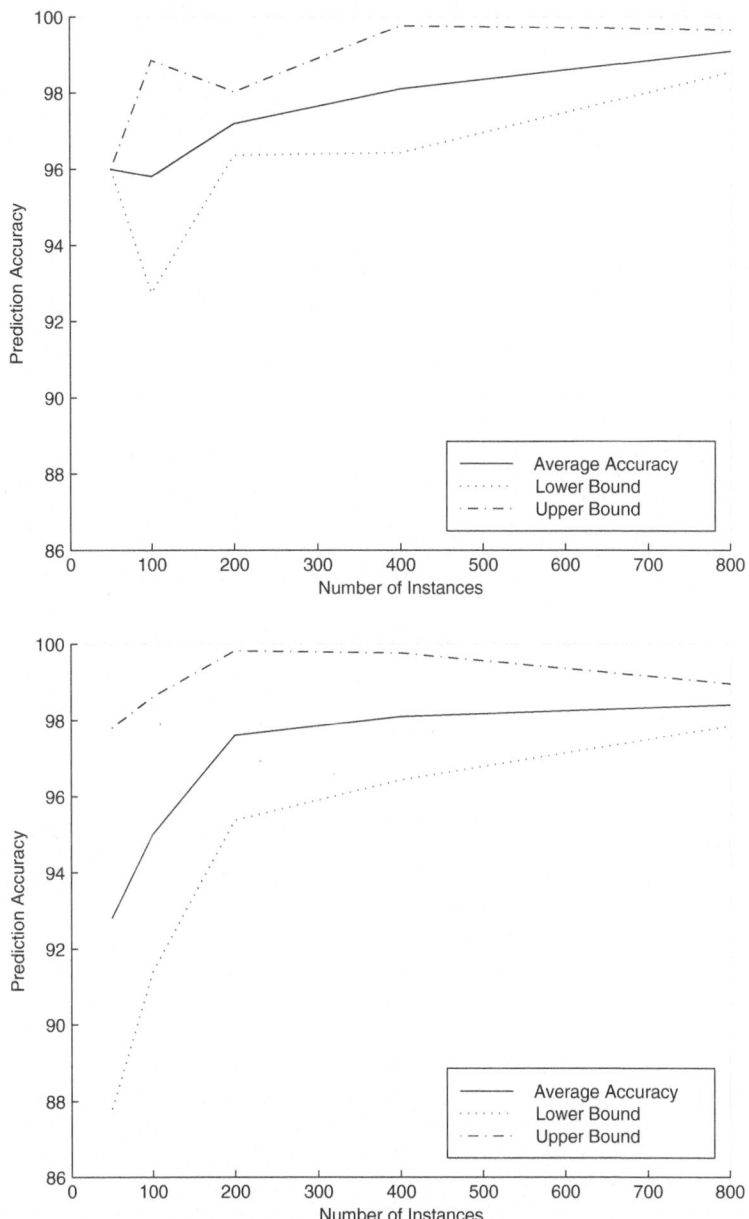

**Fig. 7.4.** Accuracy as a function of number of instances for 100 attributes (top) and 800 attributes (bottom).

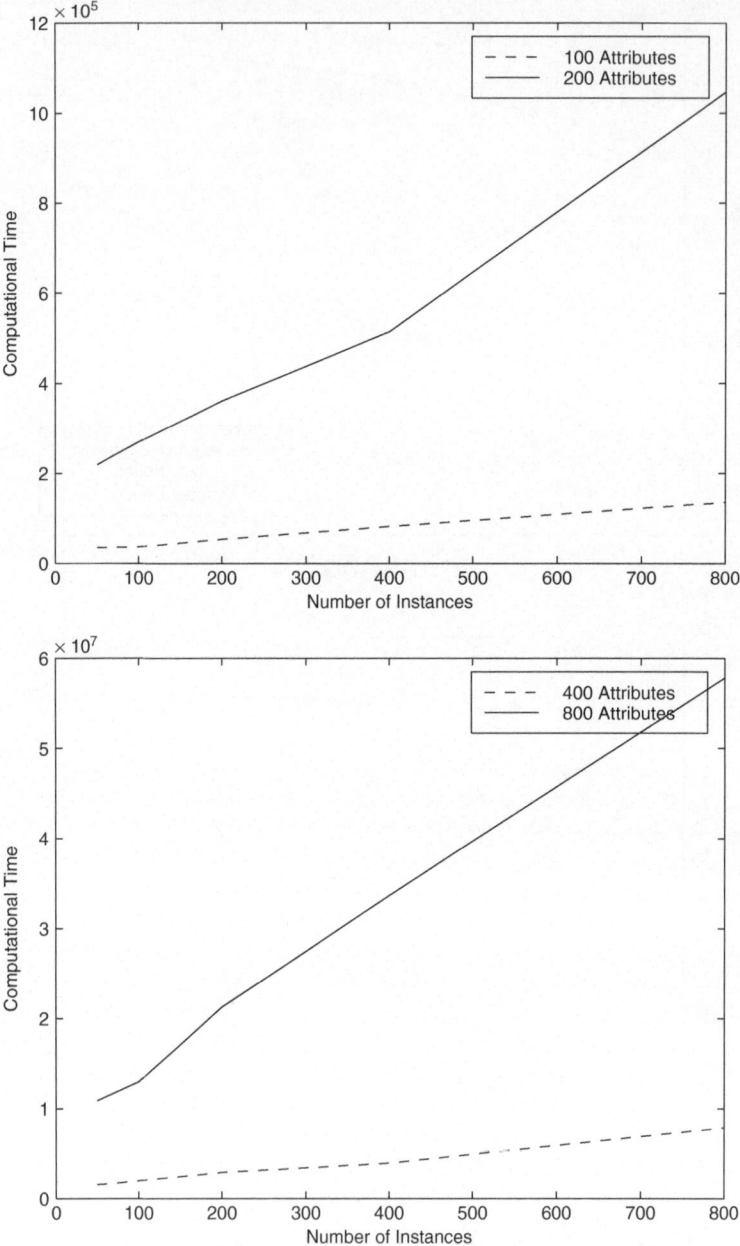

**Fig. 7.5.** Computation time as a function of number of instances.

## 7.4 Improving Efficiency through Instance Sampling

In Chapter 3 we discussed how the NP method can be applied effectively even when the objective function is noisy. This is important when applying NP for data mining problems such as the feature selection problem because in practical data mining projects it is common that the database is too large to work with all the data simultaneously. In particular, when there are millions of instances it is common to sample the database and apply the data mining algorithms to the sample instance subset. This sampling causes the objective function evaluation to be noisy since different samples of the database may result in different performance (e.g., different estimates of the accuracy of a classification model). On the other hand, using a (small) sample may be expected to greatly improve the speed of the data mining algorithm.

In light of the above discussion, we propose the following variant of the NP-Filter. In the $k$th iteration, randomly sample and use a set $\Psi_k$ of instances. To reduce bias, a new instance subset should be sampled in each iteration independently of previous sets. Thus, if the new instances indicate that an erroneous decision has been made the backtracking step of the NP method enables the algorithm to make corrections. Our main premise is therefore that the NP backtracking automatically reduces the potential bias introduced by the sampling. Several issues must be addressed regarding this approach:

- The value of random sampling in the NP-Filter must be evaluated. If the use of random sampling of instances cannot significantly reduce the computation time then the goal of increased scalability will certainly not be reached.
- The premise of automatic correction using backtracking must be assessed.
- As always when sampling of instances is used in data mining, the issue of sample size must be addressed, that is, we must determine the appropriate size of the set $\Psi_k$ of instances.

To evaluate these questions empirically, we conduct a set of numerical experiments with the NP-Filter. The datasets are the same as before, and for each dataset we run seven experiments, with 2%, 5%, 10%, 20%, 40%, 80%, or 100% of the instances used.

The results are reported as average and standard deviation over five replications, and are shown in Table 7.11. First note that the desired speedups in the algorithm are indeed achieved. By using 10% rather than 100% of the instances, the computing time is reduced by 71%, 13%, 39%, 93%, and 28% for the five datasets, respectively. This improvement is statistically significant for all but the 'audiology' dataset. Even though we accomplished the speedup, the performance as measured by the accuracy is not significantly sacrificed for any of the datasets. In the case of the 'kr-vs-kp' dataset, the accuracy even increased. These are promising results and motivate that random sampling is a reasonable way to improve the scalability of the NP-Filter. However, as illustrated by the difference in improvements between the test datasets, the

**Table 7.11.** Effect of instance sampling on speed and accuracy.

| Data Set | Fraction | Accuracy | Speed (millisec) | Backtracking |
|---|---|---|---|---|
| vote | 100% | 93.5±0.4 | 2820±93 | 0.0±0.0 |
| | 80% | 92.8±0.6 | 2766±224 | 0.0±0.0 |
| | 40% | 92.2±0.5 | 1694±352 | 0.0±0.0 |
| | 20% | 92.6±1.3 | 1065±174 | 0.6±0.5 |
| | 10% | 92.4±1.0 | 816±167 | 1.6±2.2 |
| | 5% | 91.9±1.7 | 947±515 | 13.2±18.5 |
| | 2% | 92.6±1.1 | 1314±728 | 90.4±66.7 |
| audiology | 100% | 69.7±1.9 | 41105±3255 | 0.0±0.0 |
| | 80% | 70.2±1.9 | 58230±18616 | 78.8±66.8 |
| | 40% | 70.2±2.3 | 38462±3451 | 108.6±13.3 |
| | 20% | 70.5±1.0 | 37280±26368 | 235.0±214.6 |
| | 10% | 69.2±2.4 | 35840±14563 | 371.0±182.2 |
| | 5% | 69.6±1.9 | 37025±14612 | 566.2±279.1 |
| cancer | 100% | 73.2±0.6 | 795±83 | 0.0±0.0 |
| | 80% | 73.6±0.3 | 793±26 | 0.8±1.3 |
| | 40% | 73.0±0.8 | 647±142 | 1.8±1.5 |
| | 20% | 73.3±0.8 | 640±140 | 3.8±4.1 |
| | 10% | 72.6±1.2 | 486±89 | 7.4±3.4 |
| | 5% | 73.1±0.4 | 947±456 | 78.4±48.8 |
| kr-vs-kp | 100% | 87.9±5.7 | 107467±8287 | 0.0±0.0 |
| | 80% | 87.3±7.3 | 87687±12209 | 0.0±0.0 |
| | 40% | 89.8±3.1 | 47741±4359 | 0.0±0.0 |
| | 20% | 86.1±4.4 | 19384±2727 | 0.4±0.9 |
| | 10% | 91.1±3.6 | 11482±2074 | 0.2±0.4 |
| | 5% | 89.0±1.2 | 7246±809 | 1.8±3.0 |
| | 2% | 88.6±2.3 | 7742±1892 | 25.8±15.6 |
| lymph | 100% | 83.3±1.2 | 1734±38 | 0.0±0.0 |
| | 80% | 84.2±1.3 | 2289±545 | 0.0±0.0 |
| | 40% | 84.3±2.0 | 1512±344 | 2.6±3.8 |
| | 20% | 84.7±1.0 | 1013±182 | 2.0±1.6 |
| | 10% | 84.5±1.1 | 1248±385 | 30.0±16.0 |
| | 5% | 83.1±2.1 | 28104±21543 | 2655±2122.0 |

effectiveness of this approach will in general depend on the particular dataset being analyzed.

The use of sampling speeds up the algorithm, but as pointed out above it also introduces noise into the performance estimation. Higher noise will lead to more incorrect moves in the NP-Filter, but since the overall quality of the solution, as measured by the accuracy reported in Table 7.11, does not degenerate, the NP-Filter appears able to compensate for such incorrect moves. The manner in which incorrect moves are corrected is through backtracking, which occurs when the algorithm discovers that the best solution in the current iteration is in the complimentary region. We would therefore expect more backtrackings when the noise is higher, and this is indeed supported by the numerical results reported in Table 7.11. The average number of backtracks clearly increases as the size of the instance sample sets decreases.

The numerical results reported so far indicate that random sampling is an effective approach in speeding up the NP-Filter, and that the backtracking element of the algorithm is effective in compensating for the increased noise. The final issue we consider in this section is the ideal size of the instance sample subset. From Table 7.11, of the seven sample rates used, 10% is best for the 'vote,' 'audiology,' and 'cancer' datasets, although it cannot be said that they are statistically significantly better than all others. For the other two datasets, 5% sample rate is best for 'kr-vs-kp' and 20% is best for the 'lymph' dataset. These results are appealing in that the 'lymph' data set is the smallest (only 148 instances), and the 'kr-vs-kp' is the largest (3196 instances). Thus, a smaller percentage of instances appear to be needed for the larger datasets.

Given the value of random sampling, the question arises as too how the best (or a good) instance sample rate should be determined. Furthermore, should this sample rate be constant, as in the experiment reported above, or allowed to change as the algorithm progresses? The numerical results reported above indicate that there is some relationship between problem size and a good sample rate, but another interesting observation is how it relates to the amount of backtrackings. For example, at the best sample rate of 10% for the 'vote' dataset the average number of backtracks are 1.6, but at 5% this sharply increases to 13.2. Similar patterns are observed for all the other datasets. The intuitive explanation for these observations is that although using smaller instance sets will generally speed the algorithm, once the noise increases to a certain level, excessive amount of backtracks are needed and this eventually starts to outweigh the benefits of a smaller instance subset. Although this observation does not prescribe the size of the instance sample set apriori, it suggests observable conditions under which the current size can be judged to be either too small or too large, enabling a dynamic adjustment of the instance sample rate. This will be explored further in the next section.

## 7.5 Adaptive NP-Filter

We have seen that using a random sample of instances can significantly improve the computation time of the NP-Filter but determining the best sample rate remains an open issue. In this section we propose a new approach to adaptively determine a good instance sampling rate. As described above, the number of backtrackings is one of the critical factors determining the computation time of the NP-Filter. Backtracking is necessary to correct mistakes, but when high noise causes excessive backtrackings it leads to increased computation time (see Section 2.4). Thus, we propose the following generic principles for adjusting the sampling rate:

- If relatively few backtrackings have been observed, then decrease the sample rate since using fewer samples will speed up the evaluation in each step.
- If a large number of backtrackings have been observed, then the performance estimates are too noisy and the sample rate should be increased. Thus, in each iteration, the sample rate will be either decreased or increased based on the amount of backtrackings observed.

To make this approach more specific, we define the following notation:

$$B_k = \text{Total number of backtracks made by the algorithm}$$
$$B_{k,N} = \text{Number of backtracks in the last } N \text{ iterations.}$$

The key idea of the new approach is that if the average number of backtracks observed in the past few iterations exceeds the average number of backtracks over the entire run of the algorithm, that is, $\frac{B_{k,N}}{N} > \frac{B_k}{k}$, then the noise has become too great and the sample rate should increased. Otherwise it should be decreased. To determine by how much the recent backtracking must exceed the average backtracking, we define a constant $c \geq 1$, so the condition for increasing the sample rate becomes:

$$\frac{B_{k,N}}{N} \geq c \cdot \frac{B_k}{k}. \tag{7.6}$$

Thus, the general rule employed is to decrease the number of samples used, except to prevent excessive backtrackings when detected by the rule above.

The amount of change in the sample rate also needs to be determined. We denote the sample rate in the $k$th iteration as $R_k$ and use the following equation for increasing/decreasing the number of instances

$$R_k = R_{k-1} + \frac{\Delta}{k} \tag{7.7}$$

where $\Delta$ is a predefined constant, but the amount of change decreases as the number of iterations $k$ increases. The idea is that larger changes in the sample rate will be required earlier on, but as the algorithm progresses the sample rate is allowed to settle down and converge. We call the NP feature selection method incorporating these ideas the *Adaptive NP-Filter*.

To test the effectiveness of the Adaptive NP-Filter, we evaluate it numerically on the same standard test problems as before. The parameters of the algorithm that are varied are the initial sample rate $R_0$, the constant $c$ used in equation (7.6) to compare recent backtracks to overall backtracks, and the number $N$ that defines how many steps are considered recent. The value of the initial sample rate is set as $R_0 \in \{5, 10, 20, 40, 80\}$. The constant $c$ is set to either $c = 1.0$ or $c = 1.2$. Since the number of steps required by the algorithm is proportional to the size of the problem as reflected by the depth of the partitioning tree, which in turn is equal to the number of features, $N$ is set to either 5% or 10% of the number of features and is therefore different for different test problems. Since the 'cancer' dataset has only 9 features, however, this dataset is omitted from the experiments that set $N$ equal to 5% of the features. Finally, the amount of change in the sample rate in equation (7.7) is set as $\Delta = 0.2$.

We now evaluate the Adaptive NP-Filter in terms of accuracy and computational time. As before, all averages and standard deviation results reported are based on 5 replications of the algorithm. The results are reported in Table 7.12 for $c = 1.0$ and $N$ set to 5% of features, in Table 7.13 for $c = 1.2$ and $N$ set to 5% of features, and in Table 7.14 for $c = 1.0$ and $N$ set to 10% of features. In each table, the initial instance sample rate is varied .. We observe that the accuracy is not significantly affected by most of the parameter settings, which indicates that the solution quality of the method is robust with respect to the parameter settings. However, the computational performance of the algorithm varies, and we observe:

- The speed depends on the value of $N$ that determines the number of recent steps in equation (7.6), and the difference can be as large as an order of magnitude.
- The initial instance sample rate can also greatly affect the speed.
- The speed appears to be closely related to the amount of backtracking, with excessive backtracking leading to long computation time.

**Table 7.12.** Numerical results of Adaptive NP-Filter with $c = 1.0$ and $N = 5\%$.

| Data set | $R_0$ | Accuracy | Speed | Backtracks | $R^*$ |
|---|---|---|---|---|---|
| vote | 5 | 92.5±1.5 | 1267±402 | 49.2±41.1 | 7.3±9.8 |
| | 10 | 93.1±0.4 | 4338±4368 | 303.4±363.1 | 3.7±2.1 |
| | 20 | 91.6±0.7 | 3662±3789 | 270.8±366.5 | 2.0±0.8 |
| | 40 | 93.1±1.1 | 3663±1482 | 240.4±147.7 | 1.6±0.2 |
| | 80 | 92.2±1.0 | 1067±61 | 0.6±0.9 | 16.3±0.3 |
| audiology | 5 | 71.3±2.1 | 36012±29713 | 129.8±124.2 | 29.0±6.9 |
| | 10 | 70.4±3.0 | 207481±169931 | 1357.4±1240.2 | 23.1±9.0 |
| | 20 | 70.8±1.6 | 47781±20390 | 167.6±88.9 | 35.4±8.1 |
| | 40 | 69.0±2.5 | 206110±258237 | 944.2±1206.5 | 33.8±14.0 |
| | 80 | 70.6±2.6 | 37078±20248 | 79.4±67.1 | 59.0±4.5 |
| kr-vs-kp | 5 | 86.6±2.4 | 7122±3167 | 13.8±8.9 | 4.1±4.2 |
| | 10 | 84.9±4.3 | 7122±1306 | 16.8±5.7 | 4.7±2.3 |
| | 20 | 85.5±5.0 | 9932±6555 | 17.2±9.7 | 6.0±5.7 |
| | 40 | 87.2±4.7 | 12054±7838 | 15.6±10.6 | 5.4±5.4 |
| | 80 | 83.4±6.6 | 262008±512274 | 89.2±194.5 | 27.9±16.1 |
| lymph | 5 | 84.3±1.3 | 10790±9759 | 658.2±640.0 | 10.6±0.7 |
| | 10 | 83.7±0.9 | 13800±11248 | 830.0±699.8 | 15.6±10.6 |
| | 20 | 84.3±1.2 | 5625±4392 | 318.4±282.5 | 11.8±8.0 |
| | 40 | 84.9±1.4 | 12193±16855 | 701.4±1041.6 | 7.4±3.9 |
| | 80 | 83.9±0.7 | 1107±210 | 6.8±8.0 | 12.3±1.1 |

**Table 7.13.** Numerical results of Adaptive NP-Filter with $c = 1.2$ and $N = 5\%$.

| Data set | $R_0$ | Accuracy | Speed | Backtracks | $R^*$ |
|---|---|---|---|---|---|
| vote | 5 | 92.5±1.0 | 903±587 | 19.4±27.7 | 8.6±8.8 |
| | 10 | 91.7±1.2 | 655±77 | 7.2±3.5 | 6.9±4.9 |
| | 20 | 92.2±0.5 | 663±66 | 5.8±0.8 | 5.3±0.5 |
| | 40 | 91.8±1.5 | 869±194 | 6.0±4.7 | 9.9±8.9 |
| | 80 | 93.7±0.6 | 1490±77 | 0.4±0.9 | 33.4±9.7 |
| audiology | 5 | 71.9±4.1 | 53054±14825 | 194.0±137.2 | 15.1±13.0 |
| | 10 | 69.0±3.6 | 26173±12959 | 107.0±58.1 | 31.9±9.3 |
| | 20 | 71.3±1.5 | 44677±19774 | 132.0±69.1 | 39.5±2.8 |
| | 40 | 68.3±1.5 | 31386±10880 | 89.2±46.2 | 43.0±9.5 |
| | 80 | 70.9±2.6 | 64686±32568 | 111.4±106.9 | 77.7±10.2 |
| cancer | 5 | 73.2±0.9 | 507±153 | 20.8±14.2 | 12.9±9.7 |
| | 10 | 73.4±0.7 | 444±98 | 13.4±6.3 | 14.1±2.9 |
| | 20 | 73.4±0.5 | 866±502 | 47.6±47.0 | 25.2±23.9 |
| | 40 | 73.8±0.3 | 1007±369 | 55.0±36.4 | 22.4±12.7 |
| | 80 | 74.1±0.5 | 523±32 | 0.8±0.8 | 39.5±8.4 |
| kr-vs-kp | 5 | 88.0±5.0 | 7835±3419 | 5.6±1.1 | 6.3±4.4 |
| | 10 | 86.0±4.2 | 8556±4984 | 6.4±3.8 | 6.8±5.1 |
| | 20 | 86.2±4.4 | 13301±11386 | 13.8±12.2 | 9.3±9.4 |
| | 40 | 87.0±3.9 | 15104±3659 | 5.4±2.8 | 7.6±3.3 |
| | 80 | 86.1±5.5 | 60395±13765 | 2.6±3.6 | 44.5±13.2 |
| lymph | 5 | 84.6±0.9 | 4373±3750 | 319.0±359.8 | 20.1±3.3 |
| | 10 | 83.8±1.1 | 2448±2326 | 155.8±243.3 | 17.0±4.3 |
| | 20 | 84.9±1.4 | 3205±2952 | 217.6±284.8 | 16.7±8.8 |
| | 40 | 84.1±0.9 | 999±113 | 8.0±3.4 | 16.4±5.2 |
| | 80 | 84.3±0.6 | 1240±199 | 1.6±2.3 | 37.7±6.4 |

However, despite the sensitivity to the parameters, in practice it does not appear to be difficult to recommend robust initial parameters:

- The user should avoid setting the initial sample rate very small or very large (e.g., for any problem $R_0 = 20\%$ appears to be quite robust).
- The number of recent steps should not be selected too small (e.g., for any problem, setting $N$ equal to 10% of features appears to be quite robust).

We also note that from the final sampling rate ($R^*$), that the Adaptive NP-Filter converges to is similar or a little lower than the best sampling rate found above. For example, the converged rate of 'vote' dataset ranges from approximately 5% to 10%, compared with the 10% found best for the NP-Filter with constant sampling rate (see Table 7.11). Similarly, the final sampling rate converges to between 10% and 20% in 'lymph' dataset, compared with the 20% found to be best for the constant sampling rate. An intuitive explanation is that the amount of sampling needed depends on where in the partitioning tree the current most promising region is located. At the top of the tree, it is intuitive that more accurate performance estimates are needed to select the

**Table 7.14.** Numerical results of Adaptive NP-Filter with $c = 1.0$ and $N = 10\%$.

| Data set | $R_0$ | Accuracy | Speed | Backtracks | $R^*$ |
|----------|-------|----------|-------|------------|-------|
| vote | 5 | 92.4±2.2 | 700±188 | 9.8±4.0 | 9.1±6.1 |
|  | 10 | 91.2±1.2 | 762±90 | 8.6±2.6 | 7.0±4.7 |
|  | 20 | 92.6±0.8 | 669±46 | 6.4±2.1 | 5.4±2.0 |
|  | 40 | 93.0±1.7 | 859±125 | 5.8±2.2 | 9.4±7.3 |
|  | 80 | 93.6±0.7 | 1474±17 | 0.0±0.0 | 36.5±0.0 |
| audiology | 5 | 69.9±2.6 | 53206±18499 | 138.8±26.5 | 70.8±13.6 |
|  | 10 | 72.6±2.0 | 44006±33919 | 89.4±58.2 | 54.5±6.5 |
|  | 20 | 72.7±1.6 | 49587±31218 | 98.4±57.7 | 52.5±26.5 |
|  | 40 | 71.0±2.8 | 75714±20107 | 123.2±20.8 | 85.7±10.8 |
|  | 80 | 69.6±1.0 | 229611±392499 | 434.4±802.3 | 90.2±10.8 |
| cancer | 5 | 73.0±1.6 | 779±354 | 31.4±19.5 | 25.0±12.0 |
|  | 10 | 73.6±0.8 | 630±292 | 26.0±21.0 | 17.5±10.2 |
|  | 20 | 73.7±0.6 | 502±12.7 | 13.2±7.4 | 27.1±14.4 |
|  | 40 | 73.7±0.4 | 795±488 | 34.2±37.9 | 26.0±13.3 |
|  | 80 | 73.6±0.5 | 524±37 | 3.8±3.9 | 29.4±2.1 |
| kr-vs-kp | 5 | 89.9±2.3 | 7350±2296 | 5.8±2.9 | 5.3±3.9 |
|  | 10 | 86.8±2.9 | 13222±15670 | 10.4±13.8 | 8.0±9.6 |
|  | 20 | 88.5±5.5 | 9243±2440 | 6.2±2.2 | 8.1±4.7 |
|  | 40 | 90.7±1.5 | 13312±806 | 3.6±1.1 | 6.2±3.3 |
|  | 80 | 84.1±7.2 | 51772±5016 | 0.2±0.4 | 35.9±2.3 |
| lymph | 5 | 83.9±1.1 | 7754±4051 | 604.0±341.8 | 20.1±12.6 |
|  | 10 | 84.3±0.7 | 3391±5420 | 226.0±484.7 | 18.5±4.7 |
|  | 20 | 83.9±1.8 | 2996±4202 | 171.6±360.2 | 16.0±16.2 |
|  | 40 | 84.9±1.4 | 1091±127 | 5.0±3.7 | 20.6±9.5 |
|  | 80 | 84.5±1.3 | 1101±101 | 1.6±1.7 | 38.3±4.9 |

correct move, whereas at lower depth the decision becomes easier and the algorithm is able to proceed with noisier estimates and hence a lower sampling rate.

For the next numerical results, Table 7.15 compares the Adaptive NP-Filter with both the original NP-Filter and the NP-Filter using a sampling of instances with the best constant sample rate indicated in Table 7.11. In addition to the accuracy, speed, and number of backtracks, Table 7.15 also compares the final sample rate ($R^*$) of the Adaptive NP-Filter with that of the other two algorithms. The following observations can be made from the table:

- Differences in accuracy between the algorithms are small and mostly not significant.
- The Adaptive NP-Filter provides the best overall performance in terms of speed for all the datasets while maintaining an acceptable accuracy level.
- The original NP-Filter demonstrates the worst computational performance. We finally note that determining a good constant sampling rate

**Table 7.15.** Comparison of the Adaptive NP-Filter, the constant sampling rate NP-Filter, and the Original NP-Filter.

| Data set | Approach | $R^*$ | Accuracy | Speed | Backtracks |
|---|---|---|---|---|---|
| vote | Adaptive NP | 6.9 | 91.7±1.2 | 655± 77 | 7.2±3.5 |
| | Constant NP | 10 | 92.4±1.0 | 816±167 | 1.6±2.2 |
| | Original NP | 100 | 93.5±0.4 | 2820± 93 | 0.0±0.0 |
| audiology | Adaptive NP | 31.9 | 69.0±3.6 | 26173±12959 | 107.0±58.1 |
| | Constant NP | 10 | 69.2±2.4 | 35839±14563 | 371.0±182.0 |
| | Original NP | 100 | 69.7±1.9 | 41105± 3255 | 0.0±0.0 |
| cancer | Adaptive NP | 14.1 | 73.4±0.7 | 444±98 | 13.4±6.3 |
| | Constant NP | 10 | 72.6±1.2 | 486±89 | 7.4±3.4 |
| | Original NP | 100 | 73.2±0.6 | 795±83 | 0.0±0.0 |
| kr-vs-kp | Adaptive NP | 2.4 | 88.3±1.2 | 5225±1035 | 10.4±2.9 |
| | Constant NP | 5 | 89.0±1.2 | 7246± 809 | 1.8±3.0 |
| | Original NP | 100 | 87.9±5.7 | 107467±8287 | 0.0±0.0 |
| lymph | Adaptive NP | 16.4 | 84.1±0.9 | 999±113 | 8.0±3.4 |
| | Constant NP | 20 | 84.7±1.0 | 1013±182 | 2.0±1.6 |
| | Original NP | 100 | 83.3±1.2 | 1734± 38 | 0.0±0.0 |

apriori would in general be difficult, giving the Adaptive NP-Filter another advantage.

The results reported above indicated that sampling should be done more aggressively for larger data sets (e.g., the 'kr-vs-kp' data set where a 5% sampling rate is best) than for smaller data sets (e.g., the 'lymph' data set where a 20% sampling rate is best). Table 6 also shows that the speedup of the Adaptive NP-Filter relative to the Original NP-Filter is greater for the larger data sets (e.g., 95% faster for the 'kr-vs-kp' data set versus 42% faster for the 'lymph' data set).

## 7.6 Conclusions

The feature selection problem is an important data mining problem that is inherently a combinatorial optimization problem. In this chapter we have shown that the NP method can be effective used to obtain high-quality feature subsets within a reasonable time and that by using intelligent partitioning the efficiency of the algorithm can be improved by an order of magnitude. We have also developed an adaptive version of the NP algorithm that only uses a sample of instances in each iteration and is consequently capable of scaling to large databases. This is possible because of the backtracking aspect of the NP method that allows the algorithm to recover from incorrect moves made due to decisions based on a relatively small fraction of all instances.

Our numerical results indicate that the NP method only requires using a small fraction of instances in each step to obtain good solutions, and this fraction tends to decrease as the problem size increases. However, this is application-dependent and it is non-trivial to determine apriori what the optimal instance sample rate should be for the method. The Adaptive NP feature selection method, on the other hand, is able to dynamically adjust the sample rate according to the observed frequency of backtrackings - many backtrackings imply that more instances should be used and vice versa - and thus achieve the benefits of random sampling without requiring the user to apriori determine an optimal or even good sample rate. Our numerical results show that the Adaptive NP-Filter is quite scalable and much faster than the original NP feature selection algorithm. Furthermore, the Adaptive NP-Filter obtains comparable solution quality and speeds equal to or faster than the NP using a constant sampling rate, where the sampling rate is set to the best value found. This indicates that not only does it have the benefit of automatically determining the sampling rate, but the fact that it is allowed to vary may actually improve on using the best constant sample rate.

# 8

## Supply Chain Network Design

This chapter presents the first NP hybrid utilizing mathematical programming, rather than heuristics, to generate higher-quality solutions. Specifically, it employs both approaches introduced in Section 4.4. First, it decomposes the problem into the difficult decisions, which are determined using random sampling, and the easier decisions, which are completed by obtaining exact solutions to a restricted optimization problem. Second, to generate better sample solutions to the difficult decision, a biased sampling approach is devised, which incorporates the results of solving a LP relaxation of the restricted problems. The restricted problems and relaxations are solved using CPLEX, resulting in a hybrid NP/CPLEX algorithm. Finally, the hybrid algorithm also uses the relaxed problem solution to initialize the NP search, so rather than starting with the entire search space as the most promising region, it starts with a subset that is found heuristically using those LP bounds as is discussed in Section 2.4.

The application addressed in this chapter is a location problem, which describes a wide range of problems in which the goal is to locate a set of facilities in a distribution network while satisfying the given constraints. Even the most basic location models are generally computationally intractable for large problem instances. In capacitated facility location models the goal can be the selection of a predetermined number of warehouses that minimize cost or simply finding an optimal set of warehouses to minimize the total cost. Also, the problem may be single-source, where it is required that each customer is supplied by exactly one warehouse for each product (Neebe and Rao 1983) or for all products (Hindi et al. 1998), or multiple source where splitting of the demand for a product by a customer among a number of warehouses is allowed (Bramel and Simchi-Levi 1997).

## 8.1 Multicommodity Capacitated Facility Location

### 8.1.1 Background

The literature on warehouse location problems is very extensive, dealing with
different models and different scenarios (Klose and Drexl 2003). Lee (1993)
proposed a heuristic based on Benders decomposition and Lagrangian relaxation for a multicommodity capacitated single-layer (no plants) facility location problem with a choice of facility type in which each facility type offers a
different capacity for a particular product with different fixed setup costs. The
largest test problem had 200 possible warehouse locations, 200 customers, 20
products, and 10 facility types.

However, not much work has been done on the two-layer warehouse location models (Hindi and Basta 1994). One of the classical papers in solving
locations problems is Geoffrion and Graves (1974), which solves the distribution system design problem using Benders decomposition. Their largest test
problem had 5 products, 3 plants, 67 possible warehouse locations and 127
customer zones. Selection of plant locations simultaneously with the warehouse locations has also been studied in the literature (Pirkul and Jayaraman
1996). In this chapter, we do not consider selection of plants instead, our focus
will be on the type of problems where multiple layers and multiple product
types are considered and warehouse locations are selected from a possible list
of candidates.

Recent literature on facility location models offers several additional solution techniques. These include but are not limited to branch & bound and
cutting plane algorithms (Hindi and Basta 1994, Aardal 1998), Lagrangian relaxation techniques (Klincewicz and Luss 1986, Pirkul and Jayaraman 1996,
Mazzola and Neebe 1999, Klose 2000), simulated annealing (Mathar and
Niessen, 2000), and Tabu search algorithms (Crainic et al. 1996, Delmaire
et al. 1999). Branch & Bound (BB), Branch & Cut, and Lagrangian relaxation approaches are currently the most widely applied solution techniques in
this area.

### 8.1.2 Problem Formulation

Driven by the structures and data provided by our industrial partner, and similar models introduced in the literature (Geoffrion and Graves 1974), (Pirkul
and Jayaraman 1996), (Bramel and Simchi-Levi 1997), we introduce the
following model, which considers multiple manufacturing plants and multiproduct families. The model we consider, which was given in (Bramel and
Simchi-Levi 1997), has the following features:

- A set of plants and customers are geographically located in a region.
- Each customer demands variety of products that are manufactured at
  the plants. Products are shipped from plants to warehouses and then

distributed to customers. For each shipping link and product there is a per unit shipping cost.

- Each customer's demand for each product is met from a single warehouse (single-source constraint).
- A given number of warehouses must be located among a list of potential sites. A fixed opening and operating cost must be paid for each warehouse that is opened.
- Each warehouse has a capacity not to be exceeded.

We now give some notation that will be used throughout this chapter. We let $I$ denote the set of all customers, $J$ denote the set of warehouses, $K$ denote the set of products (commodities), and $L$ denote the set of plants. We also use $I$, $J$, $K$ and $L$ to denote the cardinalities of those sets. The parameters use by the formulation are as follows:

$$c_{ljk} = \text{unit shipping cost from plant } l \text{ to warehouse } j \text{ of product } k,$$
$$d_{jik} = \text{unit shipping cost from warehouse } j \text{ to customer } i \text{ of product } k,$$
$$f_j = \text{fixed cost of opening and operating a warehouse at site } j,$$
$$w_{ik} = \text{demand of customer } i \text{ for product } k,$$
$$v_{lk} = \text{capacity of plant } l \text{ for product } k,$$
$$q_j = \text{capacity (in volume) of warehouse located at site } j,$$
$$s_k = \text{volume of one unit of product } k.$$

The decision variables for the problem are as follows:

$$y_j = \begin{cases} 1 \text{ if a warehouse is opened on site } j, \\ 0 \text{ otherwise,} \end{cases} \tag{8.1}$$

$$x_{jik} = \begin{cases} 1 \text{ if customer } i \text{ receives product } k \text{ from warehouse } j, \\ 0 \text{ otherwise,} \end{cases} \tag{8.2}$$

$$u_{ljk} = \text{amount of product } k \text{ from plant } l \text{ to warehouse } j. \tag{8.3}$$

With this notation, we can now state the problem. The objective function has three components. The first component is the transportation cost between the plants and warehouses,

$$\sum_{l \in L} \sum_{j \in J} \sum_{k \in K} c_{ljk} u_{ljk},$$

the second part of the objective function measures the transportation costs from warehouses to customers,

$$\sum_{i \in I} \sum_{j \in J} \sum_{k \in K} d_{ijk} w_{ik} x_{jik},$$

and the last term of the objective is the fixed cost of locating and operating warehouses,

$$\sum_{j \in J} f_j y_j.$$

Now turning to the constraints needed for the problem. A set of single supplier constraints guarantee that to every product/customer pair there is only one warehouse assigned:

$$\sum_{j \in J} x_{jik} = 1, \forall i \in I, k \in K.$$

Warehouse capacity constraints are needed to ensure that the total amount of products shipped from warehouses does not exceed the capacities of the warehouses and that no shipments are made through closed warehouses:

$$\sum_{i \in I} \sum_{k \in K} s_k w_{ik} x_{jik} \leq q_j y_j, \forall j \in J.$$

Conservation at warehouses constraints ensure that the amount of product arriving at a warehouse from plants equals the amount of products shipped from warehouses to customers:

$$\sum_{i \in I} w_{ik} x_{jik} = \sum_{j \in J} u_{ljk}, \forall l \in L, k \in K$$

Plant capacity constraints guarantee that for every product, the amount of that product supplied by a plant does not exceed the production capacity of the plant for that product:

$$\sum_{j \in J} u_{ljk} \leq v_{lk}, \forall l \in L, k \in K.$$

Finally, the fixed number of warehouses constraint ensures that we locate exactly $W$ warehouses:

$$\sum_{j \in J} y_j = W.$$

Taking all of the objective function components and constraints together, the Multicommodity Capacitated Facility Location Problem (MCFLP) can now be stated mathematically as follows:

$$\min_{y,x,u} \sum_{l \in L} \sum_{j \in J} \sum_{k \in K} c_{ljk} u_{ljk} + \sum_{i \in I} \sum_{j \in J} \sum_{k \in K} d_{ijk} w_{ik} x_{jik} + \sum_{j \in J} f_j y_j \quad (8.4)$$

$$\sum_{j \in J} u_{ljk} \leq v_{lk}, \forall l \in L, k \in K, \quad (8.5)$$

$$\sum_{i \in I} \sum_{k \in K} s_k w_{ik} x_{ijk} \leq q_j y_j, \forall j \in J \quad (8.6)$$

$$\sum_{i \in I} w_{ik} x_{jik} = \sum_{l \in L} u_{ljk}, \forall j \in J, k \in K, \quad (8.7)$$

$$\sum_{j \in J} y_j = W, \tag{8.8}$$

$$\sum_{j \in J} x_{jik} = 1, \forall i \in I, k \in K, \tag{8.9}$$

$$y_j, x_{jik} \in \{0, 1\}, \forall l \in L, j \in J, k \in K, i \in I, \tag{8.10}$$

$$u_{ljk} \geq 0, \forall l \in L, j \in J, k \in K. \tag{8.11}$$

Before addressing the problem using a hybrid NP algorithm, we consider how it could be solved using standard mathematical programming methods.

### 8.1.3 Mathematical Programming Solutions

In Chapter 4 we discussed how the branch and bound algorithm and its variants can be applied to large and numerically difficult mixed integer programming (MIP) problems. Recall that a MIP is essentially a linear program (LP) with integrality restrictions on one or more variables. The branch and bound algorithm guarantees optimality (assuming the problem has an optimal solution) if it is allowed to run to completion, which might require unacceptably large solution times and experience memory problems.

To solve the MCFLP problem formulated above, we used the AMPL 8.1 mathematical programming modeling language to model the MCFLP and solved it using CPLEX 8.1. The reader is referred to Mitchell (2000) for a discussion of Branch & Cut algorithms for combinatorial optimization problems. Hindi and Basta (1994) obtained feasible solutions for the multiple-source MCFLP using BB, and obtained lower bounds using Dantzig-Wolfe decomposition. Their largest test problem had 3 products, 10 plants, 15 possible warehouse locations and 30 customers.

Recall that many hard integer-programming problems can be viewed as easy problems complicated by a relatively small set of side constraints (see Section 4.1.1). Dualizing the side constraints produces a Lagrangian problem that is often easy to solve and whose optimal value is a lower bound (for minimization problems) on the optimal value of the original problem. The Lagrangian problem can thus be used in place of a linear programming relaxation to provide bounds in a branch and bound algorithm.

Beasley (1993) demonstrates the use of Lagrangian relaxation (LR) methods to solve various location problems. Here we consider two different LR approaches. We start with a commonly used approach (Pirkul and Jayaraman 1996, Bramel and Simchi-Levi 1997), namely relaxing single supplier constraints and conservation at warehouse constraints, a common practice in the literature. In particular, we relax single supplier constraints (with multipliers $\lambda_{ik}$) and conservation at warehouse constraints (with multipliers $\theta_{jk}$). The resulting problem is

$$\min_{y,x,u} \quad \sum_{l \in L} \sum_{j \in J} \sum_{k \in K} c_{ljk} u_{ljk} + \sum_{i \in I} \sum_{j \in J} \sum_{k \in K} d_{ijk} w_{ik} x_{jik} + \sum_{j \in J} f_j y_j \qquad (8.12)$$

$$+ \sum_{j \in J} \sum_{k \in K} \theta_{jk} \left( \sum_{i \in I} w_{ik} x_{jik} = \sum_{l \in L} u_{ljk} \right) + \sum_{i \in I} \sum_{k \in K} \lambda_{ik} \left( 1 - \sum_{j \in J} x_{jik} \right),$$

subject to the following constraints: plant capacity (8.5), warehouse capacity (8.6), fixed number of warehouses (8.8).

Because the relaxation of conservation of flow decouples warehouse shipments from plant shipments, this problem can be decomposed into two separate problems, a plant-warehouse subproblem and a warehouse-customer subproblem. Since there is no coupling between warehouses except for the cardinality constraint $\sum_{j \in J} y_j = W$, the preceding warehouse-customer problem can be solved by considering a set of $|J|$ single warehouse problems. Specifically, for every warehouse $j$, a knapsack problem (corresponding to $y_j = 1$) with $|I| \cdot |K|$ items is solved to obtain a value for warehouse $j$.

The optimal solution to the plant-warehouse problem is then obtained by concatenating the best $W$ solutions of these separate problems (Bramel and Simchi-Levi 1997) and closing the remaining warehouses. Then a lower bound to the original problem is given by $z_{\lambda,\theta} = z_{pw} + z_{wc} + \sum_{i \in I} \sum_{k \in K} \lambda_{ik}$, where $z_{pw}$ and $z_{wc}$ are the optimal solutions of the plant-warehouse and the warehouse-customer problems, respective. To approximately maximize $z_{\lambda,\theta}$ we use a subgradient algorithm.

Since solving the Lagrangian subproblems usually results in solutions that are not feasible for the original problem many implementations of this technique use a heuristic to obtain feasible solutions from relaxed solutions. Pirkul and Jayaraman (1996) used Lagrangian relaxation to obtain lower bounds for the MCFLP while employing a heuristic to obtain feasible solutions at every iteration. Their largest test problem had 10 products, 10 plants, 30 possible warehouse locations and 75 customers. The largest problem solved in Bramel and Simchi-Levi (1997) has 9 products, 9 plants, 32 possible warehouse locations and 144 customers. A maximum distance of 100 miles is allowed between a warehouse and a customer to have a shipping link. The following section discusses two other approaches based on stronger formulations of the basic MCFLP.

## 8.2 Hybrid NP/CPLEX for MCFLP

As we have noted throughout the book, an efficient implementation of the NP method should take advantage of the problem structure to the extent possible by incorporating this structure into the NP framework.

We have also discussed how many complex problems have aspects that are difficult, and others that can be solved more efficiently. Specifically, for the MCFLP we can think of it as consisting of three parts. First, we must decide which $W$ warehouses are opened, second, we must determine from which of the open warehouse each customer receives each product, and third, we must determine how much of each product is shipped from each plant to each open warehouse.

Now, suppose it is known which warehouses are opened, that is, the values for $y_j$, $j \in J$. Let $J^1$ denote the open warehouses, that is, $y_j = 1$ for $j \in J^1$, and $y_j = 0$ for $j \in J \setminus J^1$. Then the MCFLP problem reduces to a restricted problem:

$$\min_{x,u} \sum_{l \in L} \sum_{j \in J} \sum_{k \in K} c_{ljk} u_{ljk} + \sum_{i \in I} \sum_{j \in J} \sum_{k \in K} d_{ijk} w_{ik} x_{jik} + \sum_{j \in J^1} f_j \quad (8.13)$$

$$\sum_{j \in J} u_{ljk} \leq v_{lk}, \forall l \in L, k \in K, \quad (8.14)$$

$$\sum_{i \in I} \sum_{k \in K} s_k w_{ik} x_{jik} \leq q_j, \forall j \in J^1 \quad (8.15)$$

$$\sum_{i \in I} w_{ik} x_{jik} = \sum_{l \in L} u_{ljk}, \forall j \in J^1, k \in K, \quad (8.16)$$

$$\sum_{j \in J^1} x_{jik} = 1, \forall i \in I, k \in K, \quad (8.17)$$

$$x_{jik} \in \{0, 1\}, \forall j \in J^1, i \in I, k \in K \quad (8.18)$$

$$u_{ljk} \geq 0, \forall l \in L, j \in J^1, k \in K. \quad (8.19)$$

Note that this problem has significantly fewer binary variables. By constraint (8.9) there must be $W$ variables in $J^1$, so the total number of $x$ variables is $W \cdot |I| \cdot |K|$ in the restricted problem, versus $|J| + |J| \cdot |I| \cdot |K|$ total $y$ and $x$ variables for the MCFLP. Furthermore, for any application $W << |J|$, that is, there are many more possible warehouse locations in $J$ than the $W$ warehouses that are eventually opened. Hence, $W \cdot |I| \cdot |K| << |J| + |J| \cdot |I| \cdot |K|$. The constraints of the restricted problem are also easier. In particular, constraints (8.9) are dropped, and constraints (8.6) are significantly simplified. Finally, there are many fewer constraints in constraint set (8.7) than before ($W$ versus $|J|$ constraints). It is therefore more feasible to solve this restricted problem exactly using standard MIP solvers such as CPLEX, even for relatively large problems. Thus, we use the following approach to solve the MCFLP:

- Use random sampling to determine the warehouse solution, that is, some feasible values $\tilde{y}_j$, $j \in J$, that satisfy

$$\sum_{j \in J} \tilde{y}_j = W.$$

- For each feasible solution $(\tilde{y}_1, ..., \tilde{y}_J)$ generated by random sampling in the NP algorithm, use CPLEX to complete the solution, that is, solve a restricted optimization problem to find the optimal values $x^*_{jik}(\tilde{y}_1, ..., \tilde{y}_J)$ and $u^*_{ljk}(\tilde{y}_1, ..., \tilde{y}_J)$, given the warehouse solution.

Thus, the more difficult warehouse decisions are determined using random sampling, while the exact solution can then be completed using CPLEX for the remaining decision variables.

### 8.2.1 Partitioning

Since CPLEX is used to complete the solution given a determination of which warehouses are open, the feasible region is therefore $X = \{y \in \{0,1\}^J\}$, and the partitioning is only concerned with the decision variables $y_j$, $j \in J$. Thus, the current most promising region $\sigma(k)$ in the $k$th iteration is determined by a set of warehouses that have already been fixed to be either open, denoted $J^1$ or closed, denoted $J^0$, that is,

$$\sigma(k) = \{y \in X : y_j = 1, \forall j \in J^1, y_j = 0, \forall j \in J^0\}. \tag{8.20}$$

This most promising region is then partitioned into two subregion and the complimentary region:

$$\sigma_1(k) = (y \in \sigma(k) : \tilde{y} = 1), \tag{8.21}$$
$$\sigma_2(k) = (y \in \sigma(k) : \tilde{y} = 0), \tag{8.22}$$
$$\sigma_3(k) = X \setminus \sigma(k). \tag{8.23}$$

Thus, in $\sigma_1(k)$ it has been decided that the next warehouse $\tilde{y}$ is open, whereas in $\sigma_2(k)$ the same warehouse is closed.

The sequence in which warehouses are selected to be either opened or closed determines the quality of the partitioning, and we know that a reasonable guide is to attempt to measure the importance of variables and fix the most important variables first (see Section 2.2). This implies that a warehouse ranking is needed, which can for example be done using the following algorithm:

**Algorithm** *Warehouse Ranking*

1. Pick a warehouse $\tilde{j} \in J$ from the set of warehouses and open it. Close all other warehouses, that is,

$$y_j = \begin{cases} 1, j = \tilde{j}, \\ 0, \text{otherwise}. \end{cases}$$

   Note that since only one warehouse is allowed to be open, it may not be possible to meet demand, that is, it may not be possible to satisfy constraint (8.17) with a single warehouse. It is thus necessary to relax the $x_{\tilde{j}ik}$ variables, and replace the demand constraints (8.17) with weaker constraints.

2. Solve the linear programming relaxation of the MCFLP that requires that the total warehouse throughput equals the warehouse capacity, replacing

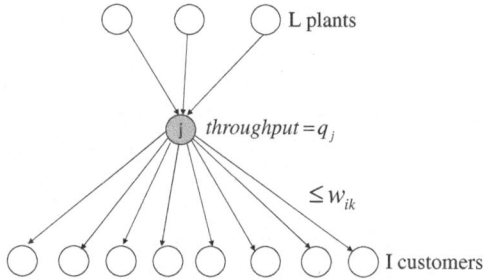

**Fig. 8.1.** Ranking heuristic iteration for warehouse $j$.

the customers' demand equations by upper bounds that limit product-customer shipments to at most the demand (see Figure 8.1):

$$\min_{x,u} \sum_{l \in L} \sum_{k \in K} c_{l\tilde{j}k} u_{l\tilde{j}k} + \sum_{i \in I} \sum_{k \in K} d_{i\tilde{j}k} w_{ik} x_{\tilde{j}ik} + f_{\tilde{j}}$$

$$u_{l\tilde{j}k} \leq v_{lk}, \forall l \in L, k \in K,$$

$$\sum_{l \in L} \sum_{k \in K} u_{l\tilde{j}k} = q_{\tilde{j}},$$

$$\sum_{l \in L} u_{l\tilde{j}k} \leq \sum_{i \in I} w_{ik}, \forall k \in K,$$

$$x_{\tilde{j}ik} \in [0,1], \forall l \in L, k \in K$$

$$u_{l\tilde{j}k} \geq 0, \forall l \in L, k \in K.$$

Note that this is a relaxed restricted problem because the integer variables corresponding to product-customer pairs are relaxed to continuous variables. The optimal value of this linear programming problem (involving only one warehouse but all plants and customers) yields the lowest total cost $\tilde{z}(\tilde{j})$ associated with full utilization of the warehouse $\tilde{j} \in J$. (Note that this optimal value also includes the fixed cost of the warehouse.)

3. Calculate an average unit cost for each warehouse by the dividing the total cost by the (fully utilized) capacity of the warehouse:

$$\bar{z}(\tilde{j}) = \frac{\tilde{z}(\tilde{j})}{q_j} \tag{8.24}$$

4. Rank the set of warehouses according to the average unit costs.

$$\bar{z}(j_{[1]}) \leq \bar{z}(j_{[2]}) \leq \dots \leq \bar{z}(j_{[J]}). \tag{8.25}$$

The ranking (8.25) can now be used to determine the sequence in which warehouses are fixed by the partitioning.

### 8.2.2 Generating Feasible Solutions

In addition to the partitioning, the other component determining the efficiency of the NP method is how successfully high-quality random solution can be generated from each region. We know that any heuristic that ranks the importance of variables can be incorporated into the NP method by using it to bias the sampling distribution used to generate feasible solutions (see Section 5.1.1). In particular, the warehouse ranking algorithm can be used for this purpose, and hence the following biased sampling algorithm can be used to generate feasible solutions from each region:

**Algorithm** *LP-Bound Sampling*

1. Given a rank cutoff value of $R$, place the top $R$ ranked warehouses according to equation (8.25) in a set called $TOP$, and the remaining warehouses are in set $BOTTOM$:

$$TOP = \left\{ j_{[1]}, ..., j_{[R]} \right\}, \tag{8.26}$$

$$BOTTOM = \left\{ j_{[R+1]}, ..., j_{[J]} \right\}. \tag{8.27}$$

2. Let $J^1 = \{j : y_j = 1\}$ denote the open warehouses, and similarly, let $J^0$ denote the warehouses that have been fixed to be closed.

   If $|J^1| = W$, stop because enough warehouses have been opened. Otherwise, open another warehouse by selecting a warehouse according to the following probability distribution:

$$P\left[\text{Select warehouse } j\right] = \begin{cases} \frac{p}{|TOP \setminus J^1|}, & \text{if } j \in TOP, \\ \frac{1-p}{|BOTTOM \setminus J^1|}, & \text{if } j \in BOTTOM. \end{cases} \tag{8.28}$$

   In other words, the probability of choosing a warehouse in the sampling scheme from $TOP$ is $p$ and the probability of choosing it from $BOTTOM$ is $1 - p$. Within each set the probability is uniform.

3. After it has been determined which warehouses will be opened, solve the restricted optimization problem defined by equations (8.13) to (8.19) to find the optimal values $x^*_{jik}$ and $u^*_{ljk}$.

We can now put the partitioning and sampling algorithms together in a hybrid NP algorithm.

### 8.2.3 Hybrid NP/CPLEX Algorithm

The hybrid NP algorithm will use the warehouse ranking (8.25) in both the partitioning and generation of feasible solutions. Note that a simple heuristic solution to the problem would open the top $W$ ranking warehouses and then complete the solution by solving the restricted problem using CPLEX. While this is unlikely to be the optimal solution, it is likely that a few of the top ranked warehouses are included in all good solutions, and in the optimal solution in particular. It may thus be possible to speed up the NP search by defining the initial most-promising region by opening some $w_0$ number of the highest-ranked warehouses.

By combining the partitioning, solution generation, and initialization procedures described above, we now obtain the following NP/CPLEX hybrid algorithm:

**Algorithm** *NP/CPLEX Hybrid*

0. **Initialization.** The initial promising region is obtained by fixing the highest ranked $w_0$ warehouses as open. In other words,

$$\sigma(0) = \left\{ y \in X : y_{[j]} = 1, j \leq w_0 \right\}, \tag{8.29}$$

where the ranking is obtained from equation (8.25) above. Set $k = 0$.

1. **Partitioning .** Let $J^1$ and $J^0$ denote the warehouses that are fixed to open and closed, respectively, in $\sigma(k)$. (For example, $J^1 = \{j_{[1]}, ..., j_{[w_0]}\}$, $J^0 = \emptyset$ for $\sigma(0)$.)

   Another warehouse $\tilde{y} \notin J^1 \cup J^0$ is selected randomly from the set $TOP \backslash (J^1 \cup J^0)$ with probability $p$, and from the set $BOTTOM \backslash (J^1 \cup J^0)$ with probability $(1 - p)$, where $TOP$ and $BOTTOM$ are defined according to (8.26) and (8.27), respectively.

   The selected warehouse $\tilde{j}$ is set to be open in subregion $\sigma_1(k)$, and closed in subregion $\sigma_2(k)$. In other words, we partition the most promising region $\sigma(k)$, into two subregions:

$$\begin{aligned} \sigma_1(k) &= (y \in \sigma(k) : \tilde{y} = 1) \\ &= \left(y : y_j = 1, \forall j \in J^1, j_j = 0, \forall j \in J^0, \tilde{y} = 1\right), \\ \sigma_2(k) &= (y \in \sigma(k) : \tilde{y} = 0) \\ &= \left(y : y_j = 1, \forall j \in J^1, j_j = 0, \forall j \in J^0, \tilde{y} = 0\right). \end{aligned}$$

   In the complementary region at least one of the warehouses used to define the promising region is not allowed in the warehouse sets considered in this region, that is,

$$\sigma_3(k) = X \setminus \sigma(k).$$

2. **Generating Feasible Solutions.** A two step process is used to generate feasible solutions. Specifically, the following steps are repeated $N_j$ times for each regions $\sigma_j(k)$, $j = 1, 2, 3$:

   a) To generate the $i$th sample point, first use random sampling to determine the values for the $y$ values not fixed by the partitioning. Specifically, the LP-Based Sampling algorithm and the biased distribution (8.28) are used to create a partial sample point $y_i^j$.

b) Second, solve the restricted problem (8.13) - (8.19) using CPLEX to obtain a completed sample point $(y_i^j, x_i^j, u_i^j)$.

After repeating the sample point generation process for each region, calculate the corresponding performance values:

$$z(y_1^j, x_1^j, u_1^j), z(y_2^j, x_2^j, u_2^j), ..., z(y_{N_j}^j, x_{N_j}^j, u_{N_j}^j), \quad j = 1, 2, 3.$$

3. **Calculate Promising Index.** For each region $\sigma_j$, $j = 1, 2, 3$, calculate the promising index as the best performance value within the region:

$$I(\sigma_j) = \min_{i=1,2,...,N_j} z(y_i^j, x_i^j, u_i^j), \quad j = 1, 2, 3.$$

4. **Move.** Calculate the index of the region with the best performance value.

$$\hat{j}_k \in \arg\min_{j=1,2,3} I(\sigma_j), \quad j = 1, 2, 3.$$

If more than one region is equally promising, the tie can be broken arbitrarily. If this index corresponds to a region that is a subregion of $\sigma(k)$, that is $\hat{j}_k \leq 2$, then let this be the most promising region in the next iteration

$$\sigma(k+1) = \sigma_{\hat{j}_k}(k)$$

Otherwise, if the index corresponds to the complimentary region, that is $\hat{j}_k = 3$, backtrack to the previous most promising region:

$$\sigma(k+1) = \sigma(k-1).$$

Note that the warehouse ranking, and hence the LP bounds found using CPLEX, are incorporated into Step 0 - Step 2, and CPLEX is further used to complete the sample generated in Step 2. On the other hand, Step 3 and Step 4 are analogous to the pure NP algorithm.

Finally, we note that even though the NP/CPLEX hybrid utilizes the results of the warehouse ranking heuristic to initialize the search process, which is found to speed the search, the algorithm nevertheless maintains a global view of the search space. Backtracking may result in some of the $w_0$ warehouses that are initially open being closed, and when generating feasible solution even low-ranked warehoused always have a positive probability of being considered.

## 8.3 Experimental Results

To illustrate the hybrid NP/CPLEX algorithm we apply it to 17 hard problem instances out of 42 test problems. These problems have various values for 5 design parameters: numbers of plants, warehouses, open warehouses, customers, and products resulting in problems of varying size and difficulty.

**Table 8.1.** Test problems suite.

| Problem # | Plants ($|L|$) | Warehouses ($|J|$) | Opened Warehouses ($W$) | Customers ($|I|$) | Products ($|K|$) |
|---|---|---|---|---|---|
| 10 | 5 | 100 | 10 | 50 | 10 |
| 12 | 5 | 100 | 10 | 200 | 10 |
| 16 | 5 | 100 | 20 | 200 | 10 |
| 20 | 10 | 30 | 10 | 200 | 10 |
| 26 | 10 | 100 | 10 | 50 | 10 |
| 27 | 10 | 100 | 10 | 200 | 3 |
| 28 | 10 | 100 | 10 | 200 | 10 |
| 32 | 10 | 100 | 20 | 200 | 10 |
| 33 | 5 | 100 | 10 | 50 | 15 |
| 34 | 5 | 100 | 10 | 250 | 5 |
| 35 | 5 | 100 | 10 | 250 | 10 |
| 36 | 5 | 100 | 20 | 250 | 10 |
| 37 | 10 | 30 | 10 | 250 | 10 |
| 38 | 10 | 100 | 10 | 50 | 15 |
| 39 | 10 | 100 | 10 | 250 | 5 |
| 40 | 10 | 100 | 10 | 250 | 15 |
| 42 | 10 | 100 | 20 | 250 | 15 |

Table 8.1 shows the values for the design parameters for the problems used in this chapter. These test problems were constructed using the distributions given in detail in Pirkul and Jayaraman (1996). After randomly generating the data for each problem we froze the data so that various solution strategies could be compared for each problem.

The algorithms were coded in AMPL 9.0 and CPLEX 9.0 was used as the branch and cut solver. The computational experiments were carried out on Sun machines with 1.2 GHz UltraSPARC-III CPU, 8 MB L2 external cache and 1 GB RAM.

The lower bounds obtained by using both LR approaches and strong MIP formulation are usually quite similar (Table 8.2). This is most likely due to the fact that we initialize the Lagrangian multipliers to be the optimal dual variables from the strong formulation LP relaxations of MDSD and the LR process generally makes only a marginal improvement to the lower bound.

Moreover, because of the effectiveness of the fast warm starting procedure employed by NP the times required by hybrid NP/CPLEX to reach highest quality solutions are generally only a few minutes (after which an even slightly better solution is rarely found). This is in strong contrast to all of the other methods which either typically produce poor quality solutions even after two hours (standard MIP approach) or which require initial solution of the relaxation of the strong MIP formulation, an initialization procedure that itself often requires an hour or more (and, in one case, more than two hours) for

**Table 8.2.** Comparison of the hybrid NP/CPLEX to other approaches for 17 difficult problems (best lower bounds in boldface).

| | Hybrid NP/CPLEX | | Traditional Lower Bounds | |
|---|---|---|---|---|
| Problem # | UB | LB | MIP LB | LR LB |
| 10 | 1,255,900 | 1,161,715 | 1,154,105 | 1,154,925 |
| 12 | 5,024,820 | 4,695,313 | 4,106,413 | 4,658,537 |
| 16 | 4,033,560 | 3,843,819 | 3,816,473 | 3,851,205 |
| 20 | 4,812,650 | 4,520,998 | 4,486,513 | 4,475,600 |
| 26 | 1,104,190 | 991,254 | 983,775 | 976,206 |
| 27 | 1,235,210 | 1,132,667 | 1,108,508 | 1,124,280 |
| 28 | 4,545,020 | 4,094,412 | 3,573,921 | 4,066,798 |
| 32 | 3,514,680 | 3,377,853 | 3,293,284 | 3,374,194 |
| 33 | 1,992,120 | 1,776,014 | 1,764,628 | 1,766,750 |
| 34 | 3,141,820 | 2,818,629 | 2,684,202 | 2,797,910 |
| 35 | 6,321,140 | 5,798,602 | 4,718,861 | 5,753,405 |
| 36 | 5,021,480 | 4,753,128 | 4,533,561 | 4,751,316 |
| 37 | 6,254,060 | 5,717,303 | 5,692,836 | 5,674,061 |
| 38 | 1,775,690 | 1,559,872 | 1,556,433 | 1,536,800 |
| 39 | 2,725,680 | 2,479,553 | 2,379,891 | 2,460,995 |
| 40 | 8,720,900 | 7,571,878 | 5,722,300 | 7,565,944 |
| 42 | 6,600,730 | 6,135,910 | 5,450,845 | 6,217,881 |

the more difficult problems. Thus, the NP framework for the use of branch-and-cut solvers to handle subproblems has proven to be efficient, reliably fast, and effective in terms of providing high-quality solutions. Specifically, for the 17 harder problems in the test suite, hybrid NP/CPLEX obtains the best solution in 14 cases, the new LR approach is better in 2 cases, while stand-alone CPLEX is better in only 1 case (corresponds to a standard MIP model).

## 8.4 Conclusions

The computational results reported in this chapter demonstrate that the NP method is capable of efficiently producing very high-quality solutions to distribution system design problems. For large-scale problems in this class, this approach is significantly faster and generates better feasible solutions than either general-purpose combinatorial optimizers (such the branch-and-cut solver within CPLEX) or specialized approaches such as those based on Lagrangian relaxation. The results illustrate that the NP framework can effectively combine problem-specific heuristics with MIP tools (such as AMPL/CPLEX), so for this problem class we have developed an excellent warehouse ranking heuristic and used this heuristic to construct a "warm start" procedure followed by an effective biased sampling approach (using CPLEX to solve the MIPs corresponding to the samples) that uses this ranking and is guided by a global view of the problem. This work has also established the applicability of the AMPL/CPLEX modeling-language/solver combination as an excellent means for the implementation of the NP method and thus represents a novel and successful use of these powerful software tools.

# 9

## Beam Angle Selection

This chapter presents an application where the evaluation of solution quality is extremely complex. The solutions, namely plans for radiation treatment, must be evaluated for their success at treating a specific patient, that is, hitting the target tumor while minimizing the harm done to vital organs. Thus, evaluating the performance may involve expert judgement from clinicians, as well as quantitative assessment by commercial software. Such external evaluation can be effectively incorporated into the promising index of the NP method (see Section 2.5), which is illustrated in this chapter. Furthermore, since this evaluation of the promising index is very expensive, only a few iterations of the NP algorithm may be possible. It thus becomes imperative to incorporate special structure into both the partitioning (see Section 2.2) and generation of samples (see Section 2.3) to ensure that the search makes rapid progress. To achieve this, this chapter considers two ways of intelligent partitioning for the problem, as well as biased sampling and heuristics for quickly generating high-quality solutions.

### 9.1 Introduction

Modern healthcare treatment technologies allow clinicians to develop complex treatment plans for a wide array of illnesses, including many forms of cancer. While expert judgment may lead to good treatment plans, as for other planning and scheduling problems the number of alternative plans usually grows exponentially and it quickly becomes impossible for an expert to identify the best plan. An alternative is to formulate the treatment planning as an optimization problem and solve for the optimal solution. As we will see in this chapter, this often yields treatments that are significantly better than those previously selected by experts.

The Beam Angle Selection (BAS) problem for Intensity Modulated Radiation Therapy (IMRT) was introduced in Chapter 1. In this chapter we explore this problem in detail and show how the NP method can be used for automating and improving the selection of the beam angles. Importantly for this application, the practice and power of current available commercial softwares and clinician's valuable experiences are easily incorporated in the NP framework. The computational results reported show that using the NP method leads to significant reductions in treatment complexity and delivery time relative to treatments generated by the commercial radiation treatment planning software currently used in clinics. We give the necessary background for the treatment planning problem that we are addressing.

### 9.1.1 Intensity-Modulated Radiation Therapy

Intensity-modulated radiation therapy is a recently developed complex technology for radiation treatment. It employs a multileaf collimator to shape the beam and to control, or modulate, the amount of radiation that is delivered from each of the delivery directions (relative to the patient). Due to the complexity of delivering IMRT, the treatment planning problem is generally divided into three subproblems. The first of these is termed the beam angle selection problem. In essence, beam angle selection requires the determination of roughly 4-9 angles from 360 possible angles subject to various spacing and opposition constraints. It is computational intense to solve this selection problem. In modern clinics the rotation angles of treatment couch is also considered as another set of decision variables. This adds even more complexity to the problem. Because of this reason, currently the angles are selected manually by clinicians based on their experiences.

The second phase of IMRT treatment planning is dose optimization. This problem includes optimizing beamlet weights as described in Langer et al. (1996), Preciado-Walters et al. (2004), and Lee, Fox and Crocker (2003). In IMRT multiple sub-apertures or segments are used at each pre-selected beam angle in order to achieve different fluence values over the grid of beamlets that comprise the overall aperture. Dose optimization optimizes the radiation delivered from each beamlet. A variety of objectives have been considered for this problem, such as minimizing a weighted sum of dose delivered to the organs-at-risk, including maximizing the minimum tumor dose. Difficult versions of this problem include dose-volume histogram (DVH) constraints, which assure that no more than $\alpha\%$ of an organ-at-risk can exceed a certain dose and at least $\beta\%$ of the tumor tissue should exceed a certain dose. Different optimization methods have been used to solve this problem, such as simulated annealing, column generation, and general mixed integer programming. For different cases and different approaches the solution time varies from minutes to hours. As oppose to solving this problem separately, we combine an easy to understand LP based dose optimization algorithm with beam angle selection using the NP method.

After applying the NP method we thus not only obtain the set of angles to deliver the radiation but also the optimized *intensity maps* for each angle as well. These intensity maps form the input for the last phase, called intensity-map segmentation or leaf sequencing (Langer, Thai and Papiez 2001). The goal is to use as few apertures as possible to deliver the optimized radiation intensity pattern to the patient. The number of segments is an important factor in treatment session time. Increasing segment counts also leads to more possibilities for delivery error due to patient motion. Currently, mixed integer programming and heuristic methods are used to solve the segmentation problem. This phase is usually solved separately from the beam angle selection and dose optimization.

### 9.1.2 Beam Angle Selection

Designing an optimal IMRT plan requires the selection of beam orientations from which radiation is delivered to the patient. These orientations, called beam angles, are currently manually selected by a clinician based on his/her judgment.

The planning process proceeds as follows: A dosimetrist selects a collection of angles and waits ten to thirty minutes while a dose pattern is calculated. The resulting treatment is likely to be unacceptable so the angles and dose constraints are adjusted and the process repeats. Finding a suitable collection of angles often takes several hours. The goal of using optimization methods to identify quality angles is to provide a better decision support system to replace this tedious repetitive process.

An integer programming model of the problem contains a large number of binary variables and the objective value of a feasible point is evaluated by solving a large, continuous optimization problem. For example, in selecting 5 to 10 angles, there are between $4.9 \times 10^{10}$ and $8.9 \times 10^{19}$ subsets of $\{0, 1, 2, ..., 359\}$. This fact has lead researchers to investigate heuristics (Ehrgott, Holder and Reese 1995).

A relatively simple heuristic for BAS is the set covering approach (Ehrgott and Johnston 2003). An angle *covers* a dose point if the sum of all the doses delivered from the beamlets of the angle to the point is greater than certain threshold. The cost of selecting an angle is large if it contains a large number of beamlets that deliver dose to a critical structure. Other researchers have heuristically solved the angle selection problem by scoring each angle and selecting a high quality set of angles (Pugachev and Xing 2001, Rowbottom, Webb and Oldham 1998, Sultan 2002, Woudstra and Storchi 2000). An angle's score increases as the beamlets that comprise the angle become capable of delivering more radiation to the target without violating the restrictions placed on the non-targeted regions. These heuristics consider beamlet individually and then aggregate this information to form a score for the entire angle. High scores are considered desirable since they indicate that it is possible to deliver large amounts of radiation to the target while maintaining the restrictions on the remaining tissues. Thus the scoring technique uses the bounds on the non-targeted tissues to form constraints, and the score represents how well the target can be treated under these constraints. This is reverse of the perspective in set covering, where the constraints attempt to guarantee that the target is treated and the objective function strives to reduce the damage to the critical structures. A vector quantization (Holder and Salter 2004) selector considers the probability that an angle is used in an optimal treatment. Although the vector quantization heuristic is deterministic, its essence is in modeling the probability that an angle is selected.

The above described heuristics reduce the search space of the original beam angle selection problem and then select a collection of angles from the reduced set, either by solving an optimization problem or through a rule. Other approaches, such as local search heuristics (Ehrgott and Johnston 2003, Das et al. 2003, Meedt, Alber and Nusslini 2003), simulated annealing (Bortfeld and Schlegel 1993, Djajaputra et al. 2003, Pugachev et al. 2001, Rowbottom, Nutting and Webb 2001, Stein et al. 1997) and genetic algorithms (Haas, Burnham and Mills 1998, Hpu et al. 2003, Schreibman 2004) have also been applied to beam angle selection problem. They search the beam space and iteratively select collections of angles, keeping track of which collection has the best objective value.

## 9.2 NP for Beam Angle Selection

We now use the beam angle selection problem to illustrate how the NP framework can be applied in radiation treatment planning problems. First we give a general formulation of the BAS problem (Ehrgott, Holder and Reese 1995). Let $A = \{0, 1, 2, ..., 359\}$ be a candidate collection of angles from which we will select $N$ to treat the patient. The goal of the objective function is to capture the criteria a treatment planner uses to decide between good and bad treatments. We let $P(A)$ to be power set of $A$ and $R_+^*$ to be the nonnegative extended reals, i.e., $R_+^* := \{x \in R : x \geq 0\} \cup \{\infty\}$, for any $A' \in P(A)$ we give the general formulation as

$$f(A') = \min_x \{z(x) : x \in X(A')\}, \tag{9.1}$$

where $X(A')$ is the feasible set of beam weights for $A'$ and $z$ maps the beam weights into $R_+^*$. If the demands of a physician cannot be satisfied with a collection of angles, i.e., $X(A') = \phi$, then we assume that $f$ assigns the value of $\infty$ to this collection. A beam weight vector $x$ represents the amount of radiation delivered from each beamlet of an angle. There are numerous approaches to estimating $f$, and the manner in which a treatment is judged varies from patient to patient and from clinic to clinic. In this chapter we will consider two approaches for evaluating beam sets. The first uses linear programming and the second uses a commercial treatment planning system called Pinnacle.

A simpler formulation, which we will use for the remainder of this chapter, is obtained by focusing only on the angles selection, rather than the weights.

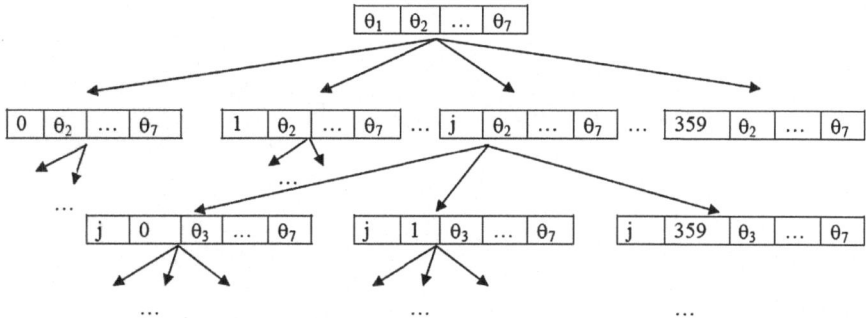

**Fig. 9.1.** Partitioning of the beam angle selection space for a 7-field plan.

Assume that we need to select $N$ angles $\theta$ to be used (with amount of radiation to be determined later), then the problem can be formulated as

$$\min_{\theta} \; f(\theta) \tag{9.2}$$

$$\text{s.t.} \;\; \theta \in X \tag{9.3}$$

$$\theta \in \{0, 1, ..., 359\}^N. \tag{9.4}$$

Here, as before, $X$ denotes the feasible region, that is, the constraints the physician(s) have specified for the set of angles. The difficult problem remains the estimation of $f$, that is, evaluating the quality of a feasible set of beam angles. We will come back to this issue in Section 9.2.3, but in the next two subsections we first address the two critical issues of partitioning and generating feasible solutions. This includes using our knowledge of the BAS to define an intelligent partitioning and using biased sampling method to obtain good feasible solutions.

### 9.2.1 Partitioning

To develop a NP algorithm for the BAS problem, first we need to consider how to partition the solution space. As we know the NP method does not limit the way in which we partition but the specific strategies employed have a high impact on the efficiency of the algorithm (see Section 2.2). For example, in Chapter 7 we saw how intelligent partitioning can improve the efficiency of the algorithm by an order of magnitude. Through partitioning, if good solutions are clustered together, the NP algorithm will then quickly identify a set of near optimal solutions.

Figure 9.2.1 shows a simple example of how to partition the solution space for the BAS problem, in which $N = 7$ (seven angles need to be selected to deliver radiation). We can first divide the solution space into 360 subregions by fixing the first index to be any one of the 360 angles. We can further partition each such subregion by fixing the second index to be any of the remaining

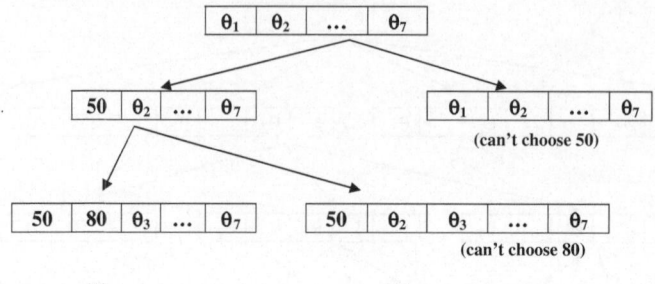

**Fig. 9.2.** Intelligent partitioning of the beam angle selection space.

angles, subject to satisfying corresponding constraints. This procedure can be repeated until a singleton region is reached, when all seven angles are selected.

The importance of intelligent partitioning that imposes good structure on the search space is discussed in Section 2.2 and demonstrated in both Chapter 7 and Chapter 8 above. We know that to impose such a structure we need a ranking of the variables, so that the most important variables can be used at the top of the partitioning tree. This tends to impose a useful structure on the search space. For the BAS problem it would be too time consuming to rank all 360 angles. Instead we note that all that is needed is to determine which angle should be used next, that is, we need a procedure to find a good angle that can then be used for the partitioning.

To find the angle to use for the partitioning we first compute the optimal solution of an integer program that optimizes beam orientation using mean organ-at-risk dose (MOD) data from single-beam plans (D'Souza, Meyer and Shi 2004). The detailed formulation is as following:

$$\min_{w} \sum_{OAR} [\alpha_{OAR}(\sum_{\theta} w_{\theta} MOD_{\theta, OAR})]$$

$$\text{s.t.} \sum_{\theta} w_{\theta} = n \tag{9.5}$$

$$w_{\theta} + w_{\theta+\delta} + w_{\theta+2\delta} + w_{\theta+(m-1)\delta} \leq 1, \theta = 0, 5, 10, ..., 355 \tag{9.6}$$

$$w_{\theta} + w_{\theta+k} \leq 1, k = 180 - \delta, 180, 180 + \delta, \theta = 0, 5, 10, ..., 355 \tag{9.7}$$

$$w_{\theta} \in \{0, 1\} \tag{9.8}$$

The objective function is the weighted average MOD over all the beams. While the beam orientation space is discretized in increments of $10°$ to generate single-beam plans within a reasonable amount of time, MOD is estimated at finer intervals of $5°$ by interpolating the MODs obtained at $10°$ increments. The $\alpha_{OAR}$ is the weight associated with an OAR, $\theta$ is the beam orientation index, $w_{\theta}$ is the binary selection variable for a beam at angle $\theta$, $MOD_{\theta, OAR}$ is the MOD for an OAR from a single beam at angle $\theta$, $n$ is number of beams to be selected, $\delta$ is the spacing between adjacent beams (in this case, $5°$) and $m\delta$ is the minimum geometric spacing required between beams. Constraint

**Table 9.1.** Illustration of frequency index.

| Frequency Index | | | | | | | |
|---|---|---|---|---|---|---|---|
| Angle | 50° 80° 110° | 250° | 280° | 310° | 350° | others |
| Index | 1   1   1 | 1 | 1 | 1 | 1 | 0 |

(9.2) specifies how many angles to select. The inequality constraints specify the minimum spacing between beams (9.3) and the exclusion of opposed (or nearly opposed) beams (9.4). This program can be solved in a few seconds using CPLEX and the solution proved to be good quality in clinical cases. But this beam angle selection problem is solely based on the mean organ-at-risk dose (MOD) information so additional data can be used to improve the solution.

Now suppose a good angle set (50°, 80°, 110°, 250°, 280°, 310°, 350°) is obtained by solving the integer program above. As shown in Figure 9.2.1 we can then partition on the first angle in the set, which is 50° in this example. Then one subregion includes angle 50°, the other excludes excluding 50°.

Similar to the initialization done in Chapter 8, we can also use the IP solution in determining the initial most promising region, that is, determine a 'warm-start' most promising region $\sigma(0)$ different from the default value of the entire feasible region. Specifically, we will do this by letting

$$\sigma(0) = \{\theta = (\theta_1, ..., \theta_2) \in X : \theta_1 = \tilde{\theta}\}, \tag{9.9}$$

where $\tilde{\theta}$ is the highest ranked angle from the IP solution, and the initial most promising region is defined by including this angle in the solution.

For instance, suppose the solution from IP is 50°, 80°, 110°, 250°, 280°, 310°, 350°, for $N = 7$. To apply intelligent partitioning, we first define a frequency index for each angle, which tells us how many times a particular angle has appeared in the best angle set. Initially the index is as shown in Table 9.1. An angle with a high frequency index means that the angle with a high probability to be a high-quality angle. Thus when defining the initial most-promising region, choose an angle with the highest frequency index to appear in the promising region. Initially all the seven angles have the same frequency index, so we could choose any one of them. For example, let angle 50° to be required in the initial most promising region, that is,

$$\sigma(0) = \{\theta \in X : \theta_1 = 50\}.$$

The frequency index is also useful in further partitioning. Since we know that the IP solution is of good quality, we want to utilize this information and pick another angle currently in the IP solution, or more specifically an angle with a high frequency index, to use for partitioning. Thus, we can for example pick angle 80° next to break the tie. That means that in the first iteration of the algorithm $\sigma(0)$ is partitioned into subregions

$$\sigma_1(0) = \{\theta \in \sigma(0) : \theta_2 = 80\},$$
$$= \{\theta \in X : \theta_1 = 50, \theta_2 = 80\},$$
$$\sigma_2(0) = \{\theta \in \sigma(0) : \theta_2 \neq 80\},$$
$$= \{\theta \in X : \theta_1 = 50, \theta_2 \neq 80\},$$

and the complimentary region is

$$\sigma_3(0) = \{w \in X : \theta_1 \neq 50\}.$$

This intelligent partitioning is illustrated in Figure 9.2.1.

Another intelligent partitioning is called a beam weight based partition. Still starting from the IP solution of 50°, 80°, 110°, 250°, 280°, 310°, 350°, for $N = 7$, we solve a dose optimization linear programming (LP) formulation that penalizes DVH violations. The LP objective is to minimize a weighted sum of doses delivered to the voxels that are over the threshold related to DVH constraints. The treatment region is divided into three regions: region $O$ includes organs-at-risk, region $T$ includes the tumor and region $N$ includes other normal tissue. The constants $b_O$, $b_T$, $b_N$ specify the thresholds for each region. The variable $x_T$ is a vector of underdoses delivered to the target region. The variable $x_O$ is a vector of overdoses delivered to the critical region. The variable $x_N$ is a vector of overdoses delivered to the normal region. The subset $sp$ of $O$, corresponds to a special region such as the spinal cord voxels. Dose delivered to $sp$ is bounded above by $b_{sp}^U$. The variable $W$ is the radiation delivered from each angle or beamlet. The overall model is thus:

$$\min \ C_T' x_T + C_O' x_O + C_N' x_N$$
$$\text{s.t.} \ A_T W + x_T \geq b_T \tag{9.10}$$
$$A_O W \leq x_O + b_O \tag{9.11}$$
$$A_N W \leq x_N + b_N \tag{9.12}$$
$$A_{sp} W \leq b_{sp}^U \tag{9.13}$$
$$W, x_T, x_O, x_N \geq 0 \tag{9.14}$$

After solving the LP dose optimization problem, we obtain the beam weights associated with the angle set as illustrated in Table 9.2. These beam weights tell us how much radiation is delivered from a particular angle in the current angle set. An angle with a high beam weight also indicates that

Table 9.2. Illustration of beam weight.

| Beam Weights | | | | | | | |
|---|---|---|---|---|---|---|---|
| Angle | 50° | 80° | 110° | 250° | 280° | 310° | 350° | others |
| Weight | 4 | 15 | 661 | 19 | 4 | 113 | 18 | 0 |

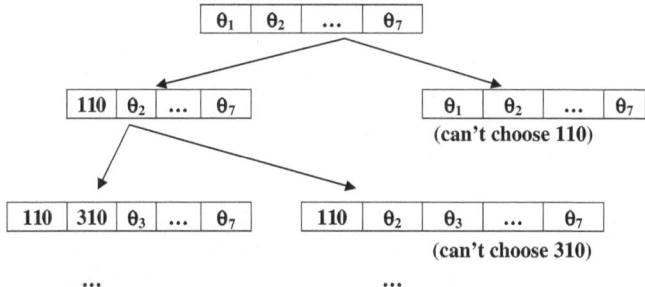

**Fig. 9.3.** Intelligent partitioning of the beam angle selection space.

the angle with a high probability to be a high-quality angle. This gives us a new way of defining both the initial promising region $\sigma(0)$ and further intelligent partitioning. In this case, we choose an angle with the highest beam weight to appear in the initial most promising region. From Table 9.2, we select angle 110° to be required in the promising region ($\theta_1 = 110$). Then the complementary region is defined as excluding angle 110° ($\theta_1 \neq 110$). This initial most-promising region definition together with further partitioning is illustrated in Figure 9.3. After defining the initial promising region, we select another angle currently in the IP solution, with second highest frequency index, to use for partitioning. Thus it is angle 310°. Then we obtain one partition $\theta_1 = 110, \theta_2 = 310$, the other is $\theta_1 = 110, \theta_2 \neq 310$. Comparing this result to Figure 9.2.1, we observe that incorporating different knowledge into the intelligent partitioning may result in completely different partitions.

### 9.2.2 Generating Feasible Solutions

Recall that the method used to randomly generate feasible sample solutions from each region in each iteration is flexible for the NP algorithm (see Section 2.3). The only requirement is that each solution point in a sampling region should have a positive probability of being selected. As we have seen throughout this book, while a uniform sampling scheme may works well in certain cases, incorporation of a simple heuristic into the sampling scheme can drastically improve the sampling quality.

A critical issue is therefore how to generate feasible samples for a dose optimization program. First, based on the angles included in the current region, we can eliminate several angles from the sample list using constraints (9.3) and (9.4). For example, for the first partition $w_{50} = 1, w_{80} = 1$, so if we let the spacing be 30°, we know that angles 25°, 30°, 35°, 40°, 45°, 55°, 60°, 65°, 70°, 75°, 85°, 90°, 95°, 100°, 105°, 225°, 230°, 235°, 255°, 260°, 265° should not be in the sample list.

After computing a list of feasible angles there are two ways to generate a sample. One way is to randomly select angles from the list with each angle in the list having equal selection probability. The other way is biased sampling.

After evaluating a sample we can obtain information related to the beam weights from the angles. Based on these weights, samples for the next iteration can be generated related to these weights. For example, angles with higher beam weights can be given higher selection probability.

After the initial sample beam sets have been obtained using the biased sampling, we can then incorporate many effective heuristic methods into the NP algorithm by using them to generate further high-quality feasible solutions from each region (see Chapter 5). In particular, heuristics often used for radiation planning can be applied effectively. This includes for example local search heuristics (Ehrgoo and Johnston 2003, Das et al. 2003, Meedt, Alber and Nusslini 2003), simulated annealing (Bortfeld and Schlegel 1993, Djajaputra et al. 2003, Pugachev et al 2001, Rowbottom, Nutting and Webb 2001, Stein et al. 1997) and genetic algorithms (Haas, Burnham and Mills 1998, Hpu et al. 2003, Schreibmann et al. 2004), each of which can be applied within each region. Since each region is a subregion of the original solution space, the heuristics can work much faster. For example, we can take the sample points as initial solutions and for each perform a fixed number of improvements based on a given heuristic method. In a parallel computing environment many such heuristics may be simultaneously applied.

### 9.2.3 Defining the Promising Index

A complication to formulating the radiation treatment planning problem as an optimization problem is that there is not a uniquely specified objective function, and in the end the plan is measured by the expert opinion of one or more physicians. This evaluation is therefore both expensive and noisy, but as is discussed in Chapter 3 such noisy performance estimates can be handled effectively by the NP method.

Thus, a variety of evaluation approaches can be used, incorporating clinical experience and knowledge, into a promising index used by the NP algorithm (see Section 2.5). For different treatment quality measures, we can use different promising index estimators. For example, we can minimize the doses to the OARs, or maximize doses to the target. We can also let the promising index to be lowest violation of the DVH constraints as well. A user-controlled interface can be set up based on these different choices. Then a user, such as a physician, can employ his/her objective and experience to define the promising index. Then NP can apply this user-selected promising index to continue partition and generating feasible solutions.

The use of such complex promising indices is demonstrated in the next section through some case studies.

## 9.3 Computational Results

In this section we present two numerical examples to illustrate the implementation and results of using the NP framework for the beam angle selection problem. One uses an LP model to evaluate beam angles generated by NP

**Table 9.3.** Data set: Pancreas case #1.

| Region | Number of voxels |
|---|---|
| Target | 1244 |
| Critical-Spinal Cord | 514 |
| Critical-Liver | 53244 |
| Critical-Left Kidney | 9406 |
| Critical-Right Kidney | 6158 |
| Normal | 747667 |
| Total | 818181 |

**Table 9.4.** Penalty values.

| | Normal | Spinal Cord | Left Kidney | Right Kidney | Liver |
|---|---|---|---|---|---|
| Penalty Values (1) | 1 | 200 | 3 | 3 | 2 |
| Penalty Values (2) | 1 | 300 | 5 | 5 | 2 |
| Threshold | - | 50% | 20% | 36% | 22% |

and the other uses a commercial available planning systems called Pinnacle to generate treatment plans from the selected beam angles.

### 9.3.1 Using LP To Evaluate NP Solutions

The first data set, also used in Olafsson and Wright (2006) and Lim (2002), is from a patient with pancreatic cancer and uses beam angle rather than beamlet data. There are several critical structures, including liver, spinal cord, left and right kidney. Distribution of voxels between the target, critical regions and normal regions is shown in Table 9.3. The full dose matrix has 36 columns (one for each 10° angle) and more than 800,000 rows (one for each voxel).

Two tests were done with different penalty values, as shown in Table 9.4 together with the threshold levels. Threshold levels are represented as percentages of the desired upper bound of each OAR.

The DVH constraints used were set as following:

- At least 96 % of the spinal cord should receive dose less than the given threshold.
- At least 60 % of the liver should receive dose less than the given threshold.
- At least 60 % of the kidneys should receive dose less than the given threshold.

Since we are solving an LP to evaluate each feasible sample, the efficiency of the LP solver is very important. Thus, a small experiment was conducted to determine a good LP strategy in CPLEX. Table 9.5 shows the test results. From these results, a good choice of strategy for this LP is primal simplex, which finishes in about 80 seconds. We therefore use the CPLEX option in bold

**Table 9.5.** LP strategy evaluation results.

| n | Method | Pricing | CPU Time (sec) | Iter |
|---|--------|---------|----------------|------|
| 1 | primal simplex | reduced | 127.98 | 42818 |
| 2 | primal simplex | combined | 118.21 | 52582 |
| **3** | **primal simplex** | **devex** | **81.91** | **43468** |
| 4 | primal simplex | st edge | >300 | - |
| 5 | primal simplex | st edge/sl | 295.49 | 52640 |
| 6 | primal simplex | full | 81.69 | 48985 |
| 7 | dual simplex | automatic | 120.14 | 29794 |
| 8 | dual simplex | standard | 111.24 | 43749 |
| 9 | dual simplex | st edge | 120.26 | 29794 |
| 10 | dual simplex | st edge/sl | 157.37 | 28609 |
| 11 | dual simplex | st edge/no | 120.53 | 29794 |
| 12 | dual simplex | devex | 88.18 | 32077 |
| 13 | barrier | Standard | 107.00 | 37429 |
| 14 | barrier | Infea. esti. | 105.62 | 37429 |
| 15 | barrier | Infea. cons. | 106.41 | 37429 |

when evaluating samples. Allowing all 36 beam angles yielded the following overdose volumes: 4.5% of cord, 20.7% of liver, 39.9% of kidneys.

Table 9.6 summarizes the results of applying the NP algorithm with the first set of penalty values. The column labeled "overdose" shows the total percentage volumn of the full set of OAR voxels that are overdosed. In this test 20 samples were generated in the promising region and 5 samples in the complementary region. This example terminated in 2 hours and 3 minutes. $L$ denotes the left "half" of the promising region and $R$ denotes the right half. In this test the left half always contained the best sample. The last four columns in Table 9.6 are the information of percentage of voxels over the threshold, which is the proper quality measure of the angle set. During the NP process the number of angles used was restricted to be less than or equal to 7. In some of the solutions, only 6 beams were used even though 7 were allowed. From the view point of DVH constraints, the solution obtained from iteration 2 (in bold) is the best.

**Table 9.6.** NP results using LP to evaluate samples.

| Iter. | Partition | No. of angles | Obj. | Normal | OAR | Overdose (%) | Spinal Cord (%) | Kidney (%) | Liver (%) |
|-------|-----------|---------------|------|--------|-----|--------------|-----------------|------------|-----------|
| 0 | - | 13 | 26478 | 20979 | 5499 | 24.9 | 4.5 | 39.9 | 20.7 |
| 1 | L | 6 | 26815 | 20964 | 5852 | 21.6 | 5.8 | 34.5 | 17.9 |
| **2** | **L** | **7** | **26809** | **21086** | **5724** | **19.6** | **3.5** | **28.2** | **17.2** |
| 3 | L | 7 | 26736 | 21049 | 5687 | 22.3 | 6.2 | 33.7 | 19.1 |
| 4 | L | 6 | 26661 | 20979 | 5682 | 26.1 | 6 | 42.5 | 21.5 |
| 5 | L | 7 | 26633 | 21017 | 5616 | 25.3 | 6.2 | 43.9 | 20 |

The details of how this NP search progressed was as follows: Solving the initial LP model, generated a solution with 13 angles and beam weights associated with each angle:

$$\begin{array}{cccccccc}
\text{Angle:} & 3 & 9 & 10 & 12 & 13 & 16 & 18 \\
\text{Weight:} & 0.003 & 0.041 & 0.092 & 0.007 & 0.014 & 0.037 & 0.095
\end{array}$$

$$\begin{array}{ccccccc}
\text{Angle:} & 19 & 24 & 27 & 29 & 34 & 35 \\
\text{Weight:} & 0.033 & 0.019 & 0.060 & 0.003 & 0.110 & 0.010
\end{array}$$

The initial most promising region was defined by including the angle with the highest beam weight from the initial LP solution. That is, angle 34 with weight 0.110, so

$$\sigma(0) = \{\theta \in X : \theta_1 = 34\}.$$

Then the partition was done on the angle with the second highest beam weight, which is angle 18 with weight 0.095. Thus, the subregions are

$$\sigma_1(0) = \{\theta \in X : \theta_1 = 34, \theta_2 = 18\},$$
$$\sigma_2(0) = \{\theta \in X : \theta_1 = 34, \theta_2 \neq 18\},$$

and the complimentary region is

$$\sigma_3(0) = \{\theta \in X : \theta_1 \neq 34\}.$$

Then sample solutions were generated by randomly selecting the other angles to make a 7-beam sample. Samples were evaluated by the same LP model. Here the best sample is the angle set $A = \{6, 10, 13, 18, 27, 34\}$, which is obtained from the partition including angle 18, that is $\sigma_1(0)$. Thus, in the next iteration angle 18 is fixed to be in the angle set, that is, $\sigma(1) = \sigma_1(0)$.

To determine which angle to use for the next partition, we use the frequency index introduced previously. Note that other than angle 34 and angle 18 that have already been fixed by the partitioning, angle 10, 13 and 27 are all the only angles included in both the initial LP-generated angle set $\{3, 9, 10, 12, 13, 16, 18, 19, 24, 27, 29, 34, 35\}$ and the best sample angle set $\{6, 10, 13, 18, 27, 34\}$ generated in the first iteration. These three angles are therefore tied with two appearances, and any of these could be used for the next level of partitioning. We randomly selected 27, and in the next iteration we thus have a most promising region

$$\sigma(1) = \{\theta \in X : \theta_1 = 34, \theta_2 = 18\},$$

subregions

$$\sigma_1(1) = \{\theta \in X : \theta_1 = 34, \theta_2 = 18, \theta_3 = 27\},$$
$$\sigma_2(1) = \{\theta \in X : \theta_1 = 34, \theta_2 = 18, \theta_3 \neq 27\},$$

**Table 9.7.** NP results using LP to evaluate samples.

| Iter. | Partition | No. of angles | Obj. | Normal | OAR | Overdose (%) | Spinal Cord (%) | Kidney (%) | Liver (%) |
|---|---|---|---|---|---|---|---|---|---|
| 0 | - | 12 | 27095 | 20973 | 6122 | 23.5 | 3.7 | 31.7 | 21.3 |
| **1** | **L** | **7** | **28292** | **21445** | **6847** | **16.1** | **2.7** | **21.7** | **14.6** |
| 2 | L | 6 | 27842 | 21240 | 6603 | 19.4 | 6.2 | 18.8 | 19.7 |
| 3 | L | 7 | 27791 | 21141 | 6650 | 21.4 | 5.6 | 31.0 | 18.7 |
| 4 | L | 7 | 27676 | 20925 | 6752 | 16.7 | 4.9 | 12.3 | 18.1 |
| 5 | R | 7 | 27661 | 21119 | 6542 | 19.2 | 3.9 | 21.5 | 18.7 |
| 6 | L | 6 | 27619 | 21147 | 6473 | 17.8 | 3.7 | 16.7 | 18.2 |

and the complimentary region is defined by

$$\sigma_3(1) = \{\theta \in X : (\theta_1, \theta_2) \neq (34, 18)\}.$$

As the procedure continued, in this case it happened that at each iteration one angle was fixed to be in the angle set, resulting in the following sequence of best angle sets:

| Iteration | Best Angle Set | Frequent Angles | Selected |
|---|---|---|---|
| 1 | 6 10 13 18 27 34 | $10, 13, 28(2)$ | $\tilde{\theta} = 27$ |
| 2 | 5 10 16 18 25 27 34 | $10(3), 13, 16(2)$ | $\tilde{\theta} = 10$ |
| 3 | 9 10 11 18 24 27 34 | $9, 11, 13, 16, 24(2)$ | $\tilde{\theta} = 24$ |
| 4 | 10 13 18 24 27 34 | $13(3), 9, 11(2)$ | $\tilde{\theta} = 13$ |
| 5 | 9 10 13 18 24 27 34 | | |

As we can see from Table 9.6, the best solution was generated in the second iteration.

For the second set of penalty values, results are shown in Table 9.7. In this test, a higher penalty was set for spinal cord region to give higher protection to it. The test finished in 1 hour and 5 minutes because we used fewer samples in the promising region.

From the results, we can see that using NP with LP to evaluate samples led to improved solutions. In particular, the solution generated via NP in the first iteration yields the following reductions in radiation relative to the original beam set: spinal cord, original overdose volume 3.7%, iteration 1 solution overdose volume 2.7% (relative reduction of 27%); kidney, original overdose volume 31.7%, iteration 1 overdose 21.7% (relative reduction of 32%); liver, original overdose volume 21.3%, iteration 1 overdose 14.6% (relative reduction of 31%). We also have more flexibility in selection of the solutions. Because we can vary the penalties for the OARs depending on the DVH constraints, we can obtain different solutions. We can also select different results within all the NP iterations (DVH effects versus objective values).

## 9.3.2 Using Condor for Parallel Sample Evaluation

An important feature of the NP method is that it is naturally parallelized and this opens up an intriguing possibility of integrating the NP method with a high-performance distributed computation environment such as Condor. Figure 9.4 illustrates the role of Condor in a distributed implementation of the NP method. In this example, we use the modeling program GAMS. An advantage of using GAMS is that it is designed for optimization and it uses powerful commercial solvers, such as CPLEX. We can generate a good initial solution via IP to start NP. Another advantage of using GAMS is that LP can be utilized to determine an index. A disadvantage is the time spent on generation of the samples. After partitioning and sampling, the samples are submitted to Condor (Litzkow, Livny and Mutka, 1988). We take advantage of the High Performance Computing Grids recently provided by GAMS. This extends the GAMS language to support asynchronous submission and collection of model solution tasks to take advantage of grid environments. Condor will treat the GAMS main program as master, then will use the script file automatically generated by the GAMS grid computation interface to distribute the jobs to the "workers", which are the available machines on the network. Our program will check for the completion of evaluation of each job. Here each job generates the promising index of the corresponding sample. When all the jobs are completed, we obtained the updated promising index for each region. The partitioning and sampling can be repeated in the same fashion. Computational results shown below demonstrate parallel evaluation of samples saves time compared to sequential evaluation.

Results using the sequential evaluation method to evaluate all the samples generated by the NP are shown in previous section. That is, we generate the first sample in the first region and use LP to obtain a "score" for this sample. Then we generate the next sample and obtain the "score". After all samples in the region are generated and evaluated, the promising index for this region is the best "score" of the samples. Then we apply the same procedure to the rest of the regions. When all the promising indices are obtained, we can define the most promising region for the next iteration. In our test using the sequential evaluation method, we generated 20 samples for the promising region and 5 samples for the complimentary region. So in each iteration of NP, a total of 25 samples were evaluated. Our program terminated after 5 iterations in 123 minutes.

The parallel evaluation was done using Condor as shown in Figure 9.4. The summary of the results for each step is shown in Table 9.8. From this table we can see that using parallel evaluation of samples, the computational time is 60 minutes, including about 9 minutes spent on waiting for the acquisition of resources. Compared to doing this sequentially, we save more than half of the computational time.

We can further reduce the computational effort by sampling normal tissues. Two more tests were done with different number of normal voxels, as shown

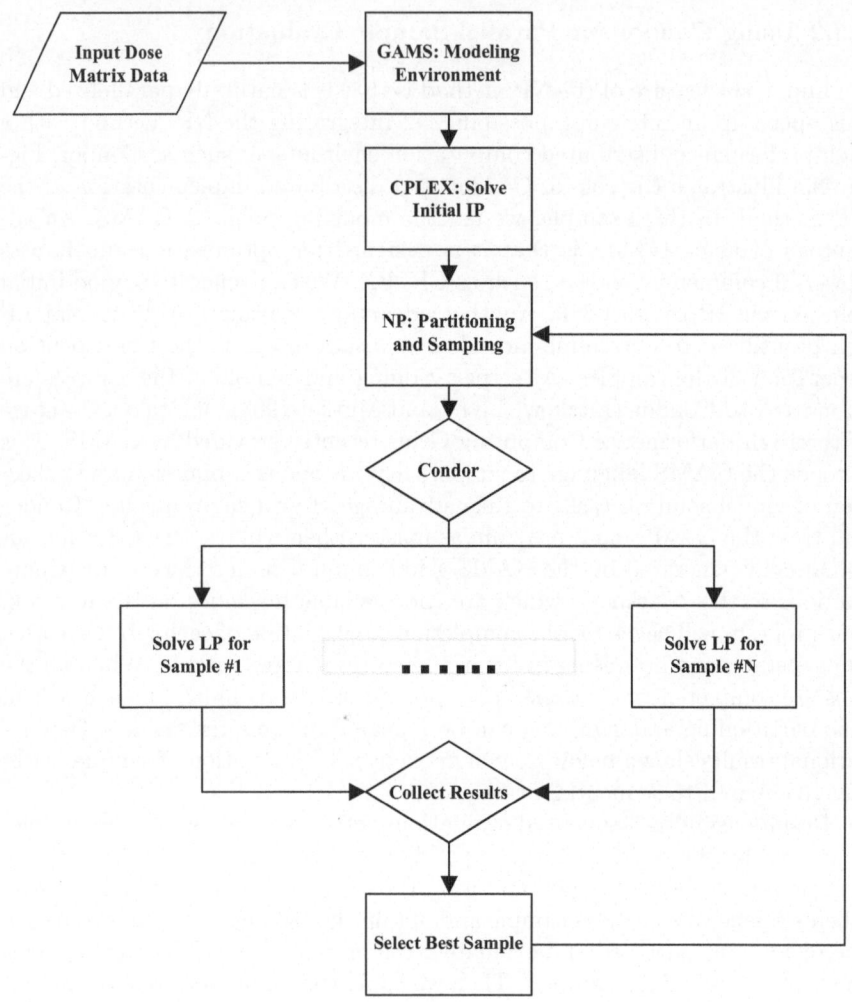

**Fig. 9.4.** Flowchart of parallel evaluation of NP samples.

in Table 9.9. They represent the full data set, approximately 10% of normal voxels and 5% of normal voxles, respectively. For the two reduced data sets, the normal voxels were sampled randomly. Test 1 terminated in 2 hours and 3 minutes without using Condor. If Condor is utilized to evaluate samples in parallel, Test 1 terminated in 51 minutes. Table 9.9 also shows the run time comparison of single NP iteration with Condor for three tests. We can see that with reduced number of normal voxels we could save about 5 minutes for each NP iteration.

**Table 9.8.** Using Condor for parallel evaluation of NP samples (5 NP iterations).

| Iteration | Task Accomplished | Time (minutes) |
|---|---|---|
| | Started program | 0 |
| Iteration 0 | Obtained initial IP solution | 4 |
| Iteration 1 | Submitted 1st sample | 4 |
| | 1st sample evaluated | 6 |
| | Submitted 25th sample | 10 |
| | 25th sample evaluated | 18 |
| Iteration 2 | Submitted 1st sample | 18 |
| | Obtained available machine | 24 |
| | 1st sample evaluated | 25 |
| | Submitted 25th sample | 25 |
| | 25th sample evaluated | 28 |
| Iteration 3 | Submitted 1st sample | 28 |
| | 1st sample evaluated | 29 |
| | Submitted 25th sample | 36 |
| | 25th sample evaluated | 38 |
| Iteration 4 | Submitted 1st sample | 38 |
| | 1st sample evaluated | 41 |
| | Obtained available machine | 44 |
| | Submitted 25th sample | 45 |
| | 25th sample evaluated | 50 |
| Iteration 5 | Submitted 1st sample | 51 |
| | 1st sample evaluated | 52 |
| | Submitted 25th sample | 58 |
| | 25th sample evaluated | 60 |

**Table 9.9.** Comparison of Condor run time for three voxel sets (single iteration).

| Time (min:sec) | Test 1 $(15^*)$ | Test 1 $(25^*)$ | Test 2 $(15^*)$ | Test 2 $(25^*)$ | Test 3 $(15^*)$ | Test 3 $(25^*)$ |
|---|---|---|---|---|---|---|
| Voxels | 747,667 (full) | | 70,000 (10%) | | 35,000 (5%) | |
| Run total | 7:22 | 11:23 | 5:00 | 6:28 | 4:22 | 6:15 |
| Generation | 5:12 | 7:37 | 2:38 | 3:13 | 1:57 | 3:07 |
| Solving | 2:10 | 3:46 | 2:22 | 3:15 | 2:25 | 3:08 |
| Acquiring | 7:23 | 1:20 | 1:00 | 3:40 | 5:23 | 1:35 |

$^*$ *Number of samples.*

### 9.3.3 Using Pinnacle To Evaluate NP Samples

As a benchmark for using the LP to evaluate samples, we also use a promising index computed via the Pinnacle system. Two cases were used, one is a head and neck case, the other is a pancreas case. The data are listed in Table 9.10 and 9.11.

**Table 9.10.** Data set: Head and neck case.

| Region | Number of voxels |
|---|---|
| PTV | 3034 |
| GTV | 849 |
| Spinal Cord | 202 |
| Left Parotid | 407 |
| Right Parotid | 614 |
| Normal | 329424 |
| Total | 334530 |

**Table 9.11.** Data set: Pancreas case #2.

| Region | Number of voxels |
|---|---|
| PTV | 10874 |
| CTV | 6934 |
| Spinal Cord | 424 |
| Left Kidney | 2886 |
| Right Kidney | 2906 |
| Liver | 30342 |
| Normal | 448297 |
| Total | 502663 |

This time the initial solution was obtained by using the IP model described in Section 9.2.1, which is solely based on MOD information. The promising region was defined by selecting an angle in the initial solution. Then the partition was based on the frequency index. We ran five iterations for each case. There are 42 samples better than the initial IP solution for the head and neck case, and 7 samples better for the pancreas case. The sample evaluator or scor-

**Table 9.12.** Head and neck case solution (Pinnacle).

| Initial Solution | Score |
|---|---|
| angle set 10  40  110 140 210 300 330 | 0.404307 |
| **Frequency Based Partition** | |
| Iteration 1 | Score |
| angle set 25  70  110 140 185 300 330 | 0.624833 |
| Iteration 2 | Score |
| angle set  5  55  110 140 205 300 330 | 0.897575 |
| Iteration 3 | Score |
| angle set  0  30  80  140 195 300 330 | 0.344441 |
| Iteration 4 | Score |
| angle set 15  50  140 180 210 300 330 | 0.280254 |
| Iteration 5 | Score |
| angle set 20 105 140 180 210 300 330 | 0.524283 |
| **Beam Weight Based Partition** | |
| Iteration 1 | Score |
| angle set 35  70  110 180 230 300 330 | 0.881381 |
| Iteration 2 | Score |
| angle set 20  50  110 175 210 300 330 | 0.605615 |
| Iteration 3 | Score |
| angle set 10  80  110 175 210 300 330 | 0.852836 |
| Iteration 4 | Score |
| angle set 20  50  135 165 210 300 330 | 0.325324 |
| Iteration 5 | Score |
| angle set 20  80  115 175 220 310 340 | 0.688058 |

**Table 9.13.** Pancreas case solution summary (Pinnacle).

| Initial Solution | Score |
|---|---|
| angle set 30 60 90 260 290 320 350 | 0.672251 |

| Iteration 1 | Score |
|---|---|
| angle set 30 60 90 260 290 320 350 | 0.672251 |

| Iteration 2 | Score |
|---|---|
| angle set 30 90 125 155 220 315 350 | 0.624467 |

| Iteration 3 | Score |
|---|---|
| angle set 30 60 90 150 255 295 350 | 0.61254 |

| Iteration 4 | Score |
|---|---|
| angle set 30 90 120 155 185 230 350 | 0.442872 |

| Iteration 5 | Score |
|---|---|
| angle set 30 90 120 155 185 230 350 | 0.442872 |

**Table 9.14.** Comparison between using Pinnacle and LP to evaluate samples.

|  | Data used | Min. Score | Ave. Score |
|---|---|---|---|
| Pinnacle | Beamlet | 0.280254 | 0.60246 |
| LP | Beamlet | 0.806525 | 1.351387 |
| IP | MOD | 0.404307 | 0.404307 |

ing scheme we used is the weighted sum of the violation of DVH constraints. Each sample generated by NP framework was input to Pinnacle to generate a full clinical plan. From the plan we obtained the percentage of voxels violating the DVH constraints and multiplied this percentage by the same weight used in the IP program described previously for each OAR. Adding the number for all the OARs we obtained the final score. The best solution obtained by NP is found to be much better than the one IP obtained, as can be seen from Table 9.12 and 9.13. Finally, Table 9.14 compares the method of using LP and Pinnacle to evaluate NP samples.

## 9.4 Conclusions

We have demonstrated that the NP method provides an effective framework for obtaining high-quality solutions to the beam angle selection problem in Intensity-Modulated Radiation Therapy. Relative to good quality beam angle sets constructed via expert clinical judgement and other approaches, the beam sets generated via NP showed significant reduction (up to 32%) in radiation delivered to non-cancerous organs-at-risk near the tumors. Thus, in addition to providing a method for automating beam angle selection, the NP framework yields higher quality beam sets that significantly reduce radiation damage to critical organs.

# Local Pickup and Delivery Problem

In Chapter 4 we introduced some general techniques for using the solutions to linear programming relaxations, as well as other mathematical programming (MP) methods, to define both intelligent partitioning and methods for generating more high-quality sample solutions. This results in what we refer to as hybrid NP/MP algorithms. In this chapter, we illustrate these techniques for a specific application, namely the local pickup and delivery problem (LPDP).

## 10.1 Introduction

In recent years, the competition in the transportation and logistics sector has become increasingly intensified. To respond to this challenge, commercial carriers have been investing heavily in new technologies and have been focusing on developing cost-cutting strategies in order to improve their profitability. For instance, the local pickup and delivery problem (LPDP), a variant of the vehicle routing problem (VRP), has drawn a great deal of interest lately. In this chapter, we aim to provide a new general solution approach for solving this type of problem.

The LPDP is concerned with the optimal movement of a set of loads in a local service area over a relatively short planning horizon. The basic operations involved in LPDP can be described as follow: At the beginning of each work day, a fixed number of vehicles are positioned throughout the service area. A vehicle can serve only one load at a time. After the delivery of a load, it runs for another load immediately or becomes idle. Served loads generate revenues and unserved ones may be subcontracted to other carriers (for a nominal fee) or simply lost (without generating any revenue). Empty movements of vehicles incur costs. The optimization objective is to maximize the overall profit over a fixed planning horizon, e.g., from the decision epoch to the end of the day. To achieve this objective, a carrier must balance between serving as many loads as possible and minimizing empty movements. Such a problem arises when small local logistics companies with fleets consisting of dozens of vehicles try to meet demand within the vicinity of a city (e.g., local taxicab companies). Another example is for some large truckload carriers with dedicated truck fleets to certain geographical regions or hubs to handle local loads. While some previous work has been concerned with VRP in dynamic and stochastic settings (Powell 2003, Gendreau, Laporte and Sequin 1996), in this chapter we deal with the static and deterministic version of the problem. The planning horizon of the LPDP is generally short, and in many applications, a deterministic model (run either in a static or rolling horizon manner) can be satisfactory.

Modeling issues relevant to LPDP have been discussed by many researchers. One set of constraints represents load-specific requirements. Time window constraints (or sometimes, pickup time window constraints) are one of the most important attributes of loads, and have been considered in various formulations (Wang and Regan 2002, Dumas, Esrosiers and Soumis 1991, Brasy and Gendreau 2005, Desrosiers et al. 1986). As often considered in applications where loads are associated with intermodal routes, service time window constraints enforce that each load either will be served within a given time window or will not be served at all. Such constraints result in significant computational difficulties when solving LPDP. Some other constraints on the load side are also considered, such as job precedence constraints (Fagerholt and Christiansen 2000) and nested precedence constraints (Xu et al. 2001).

Besides load constraints, there are also constraints imposed on the resource/vehicle, e.g., capacity and working hours of vehicles (Mourkousis, Protonotarios and Varvarigou 2003), different vehicle speeds and unit operating costs, and minimum workload of each vehicle per day (Lim, Wang and Xu 2006). Also, for years truckload carriers have been facing increasingly serious driver shortage problems, and the current turnover rate of drivers is very high (it is not rare to see 100% annually). As a result, companies are forced to pay more and more attention to keeping their drivers happy in order to combat driver attrition. Our formulation is motivated by such a consideration. Specifically, in this chapter we account for two main sets of constraints along with hard time window constraints:

- Homing driver constraints: As discussed in Pan, Shi and Pi (2005), the most important consideration in creating driver satisfaction in the planning process is to allow a driver to return home each day, should this be the driver's preference. Creating a personalized, pre-determined work schedule for the driver will clearly make the driver's life easier.
- Driver qualifications and preference constraints. For some special loads, such as just-in-time loads, they can only be served by qualified drivers. Also, drivers may have preference over types of loads, which should be accommodated whenever possible.

For both types of constraints mentioned above, we model them as hard constraints. In practice, however, it might be desirable to relax them by way of penalty terms added to the objective function. Overall, we end up with a LPDP with time window constraints and nonhomogeneous resources. In this chapter, we present a mixed integer programming (MIP) formulation for this problem.

In standard solution approaches for this type of problems, one tries to solve the problem optimally using MIP solvers (Wolsey 1998), dynamic programming (Thomas 1976), or some specialized algorithms (Arunapuram, Mathur and Solow 2003, Lu and Dessouky 2004). These methods generally are not capable of handling large-scale problems in the real world. On the other hand, fast and good approximate methods are more useful in solving such problems. In Wang and Regan (2002), the authors provided the time window reduction and partitioning method, which handles the difficulty associated with the service time window constraints. In Powell and Carvelho (1998) and Powell, Shapiro and Simao (2002), the approximate dynamic programming (ADP) method is provided, based on establishing value functions in each stage to reduce the problem size. These two methods are more efficient for problems with homogenous resources. There are also some good computational results on this type of problems through Column Generation (CG) method (Xu et al. 2001). Mostly, a large set of columns need to be generated, and the highly specialized and efficient algorithm for solving the pricing problem is the essential part of the CG procedure. Many of the heuristics proposed for the problem, such as dispatching rules, are fast, but the solution quality is often not good. Some others methods, e.g., tabu search, genetic algorithm, etc, are efficient if well designed, but highly problem-dependent.

## 10.2 LPDP Formulation

We now formulate the local pickup and deliver problem (LPDP) precisely. Recall that in this problem there is a fixed set $K$ of vehicles positioned in a service area where a set $L$ of loads needs to be served. Each vehicle can only serve one load at a time and after each load it either goes directly to serve another load or it becomes idle.

The decision variables determine both the assignment of vehicles to loads and the sequence. Specifically,

$$x_{klj} = \begin{cases} 1, & \text{if vehicle } k \text{ serves load } l \text{ and then goes on to serve load } j, \\ 0, & \text{otherwise.} \end{cases}$$

(10.1)

With these decision variables the objective function, that is, the total revenue minus the total transportation cost, can be formulated as follows:

$$z = \max_{x_{klj}} \sum_{l \in L} r_l \cdot \sum_{k \in K} \sum_{i \in L \cup \{k+|L|\}} x_{kil}$$

$$- \sum_{l \in L \cup \{k+|L|\}} \sum_{j \in L \cup \{k+|L|+|K|\}} C_{lj} \cdot \sum_{k \in K} x_{kil},$$

where $r_l$ is the net revenue of load $l \in L$ and $C_{lj}$ is the cost of traveling from the destination of load $l \in L$ to the origin of load $j \in L$.

The first set of constraints are the standard multi-commodity network flow constraints:

$$\sum_{j \in L \cup \{l+|K|\}} x_{(l-|L|)l_j} = 1, \forall l \in S,$$

$$\sum_{l \in L \cup \{j-|K|\}} x_{(l-|L|-|K|)l_j} = 1, \forall j \in H,$$

$$\sum_{i \in L \cup \{k+|L|\}} x_{kij} = \sum_{l \in L \cup \{k+|L|+|K|\}} x_{kjl}, \forall k \in K, j \in L.$$

These constrains simply assure the proper balance in the network, e.g., a truck cannot leave a node that it does not enter. The next set of constraints assures that each load is served at most once:

$$\sum_{k \in K, i \in L \cup \{k+|L|\}} x_{kij} \leq 1, \forall j \in L.$$

We note that a load does not have to be served but since the objective is to maximize the revenue of loads minus the cost, the optimization program attempts to serve all load as profitably as possible.

There may be also be some limitation on driver qualification and/or their preferences. This excludes some assignments and these are represented in the set $Q_{kl}$ for each vehicle/driver $k \in K$ and load $l \in L$, that is, we have the following constraint:

$$\sum_{j \in L \cup \{k+|L|+|K|\}} x_{kij} = 0, \forall k \in K, l \in L : Q_{kl} = 0.$$

The next set of constraints preserves the temporal relations between consecutive nodes:

$$t_l + T_{lj} - t_j \leq \left(1 - \sum_{k \in K} x_{klj}\right) \cdot b_l + T_{lj} - a_j, \forall l \in L, j \in L,$$

$$t_l + T_{lj} - t_j \leq \left(1 - x_{(1-|L|)l_j}\right) \cdot b_l + T_{lj} - a_j, \forall l \in S, j \in L,$$

$$t_l + T_{lj} - t_j \leq \left(1 - x_{(1-|L|-|K|)l_j}\right) \cdot b_l + T_{lj} - a_j, \forall l \in L, j \in H.$$

Finally, there are time windows $[a_l, b_l]$ for the pickup times of each load $l \in L$, that is,

$$t_l \geq a_l, \forall l \in L \cup S,$$

$$t_l \leq b_l, \forall l \in L \cup H.$$

In the next section we discuss an efficient and effective implementation of the NP method to solve this very complex MIP. In particular, we show

how mathematical programming techniques can be incorporated into the NP framework at various stages to improve the efficiency of the resulting hybrid NP algorithm.

## 10.3 NP Method for LPDP

We now show how an efficient hybrid NP algorithm can be developed for this problem, by incorporating exact mathematical programming methods into both the partitioning and the generation of feasible sample solution. We refer to the resulting algorithm as a hybrid NP/MP algorithm.

### 10.3.1 Intelligent Partitioning

To develop an intelligent partitioning method, we note that the decision variables contain two separate decisions: the assignment of vehicles to loads and the sequence of loads assigned to the same vehicle. To separate the two we rewrite the decision variables as

$$x_{klj} = y_{kl} \cdot y_{kj} \cdot s_{lj}, \tag{10.2}$$

where

$$y_{kl} = \begin{cases} 1 \text{ if vehicle } k \text{ is assigned to load } l, \\ 0 \text{ otherwise}, \end{cases} \tag{10.3}$$

$$s_{lj} = \begin{cases} 1 \text{ if load } l \text{ is directly followed by load } j. \\ 0 \text{ otherwise}. \end{cases} \tag{10.4}$$

We have now decoupled the two decisions, which is important because it turns out that once assignments are made the remaining sequencing decision can be solved fairly easily using mathematical programming methods. As was pointed out in both Chapter 1 and Chapter 4, the ability to do this type of decoupling is fairly common for many complex integer programming problems, and the NP method can take advantages of this in an effective manner.

As we did in Chapter 4, we will take advantage of the flexibility of the NP method by using random sampling to obtain solutions to the more difficult assignment decisions ($y = \{y_{kl}\}$), and then solve an IP to complete the solution, that is, obtain the values for the sequencing decisions ($s = \{s_{lj}\}$), given the assignments. We note that in general sequencing problems are hard but the structure and moderate size of the sequencing problem considered here make them tractable.

The solution space to be partitioned thus becomes:

$$X = \left\{ y : \sum_{k \in K} y_{kl} = 1, \forall l \in L; y_{kl} \in \{0, 1\}, \forall k \in K, l \in L \right\}. \tag{10.5}$$

We note that this is simply a binary integer program (BIP) with $|K| \times |L|$ zero-one decision variables, and a constraint that assures that each load is assigned to exactly one vehicle.

We now develop an intelligent partitioning procedure for the assignment decisions. As we have done in previous applications (see e.g., Chapter 7 and Chapter 9), we need to sequence the variables in order of importance, so the most important variable can be used first in the partitioning. Since this is a BIP, equation (4.18) from Chapter 4 could be applied to define a sequence among these variables. However, (4.18) has a symmetry between variables that are set to zero and variables that are set to one. For the LPDP almost all of the variables are set to zero so it is the ones that are set to one that are of the most importance. We therefore solve the LP relaxation and define a new sequence based on the relaxed solution $y^{LP} = \{y_{kl}^{LP}\}_{k \in K, l \in L}$ such that

$$z\left(y_{(k,l)[1]}^{LP}\right) \geq z\left(y_{(k,l)[2]}^{LP}\right) \geq \ldots \geq z\left(y_{(k,l)[n]}^{LP}\right), \tag{10.6}$$

where $n = |K| \cdot |L|$ is the total number of variables. We can then proceed by first partitioning based on $y_{(k,l)[1]}^{LP}$ being either zero or one, then $y_{(k,l)[2]}^{LP}$, and so forth. Specifically, in the first iteration the most-promising region will be defined as $\sigma(0) = X$, and this is then partitioned into two subregions:

$$\sigma_1(0) = \left\{y \in X : y_{(k,l)[1]}^{LP} = 1\right\}, \tag{10.7}$$

$$\sigma_2(0) = \left\{y \in X : y_{(k,l)[1]}^{LP} = 0\right\}. \tag{10.8}$$

This partitioning is then continued in the same manner.

Say, for example, that we have a problem with two vehicles ($|K| = 2$) and seven loads ($|L| = 7$). We solve the LP relaxation, and obtain the following solution:

$$y_{kl}^{LP} = \begin{cases} 0.10, & \text{if } k = 1, l = 1, \\ 0.50 & \text{if } k = 1, l = 2, \\ 0.20 & \text{if } k = 1, l = 3, \\ 0.10 & \text{if } k = 2, l = 5, \\ 0.05 & \text{if } k = 2, l = 6, \\ 0 & \text{otherwise.} \end{cases} \tag{10.9}$$

We now have a ranked list of assignments that can be used to define an intelligent partitioning:

$$y_{12}^{LP} > y_{13}^{LP} > y_{11}^{LP} = y_{25}^{LP} > y_{26}^{LP}.$$

This means that since $y_{12}^{LP} = 0.50$ takes the highest value, the assignment of Load 2 to Vehicle 1 can be considered the most critical, and the top level partitioning is:

$$\sigma_1(0) = \{y \in X : y_{21} = 1\},$$
$$\sigma_2(0) = \{y \in X : y_{21} = 0\}.$$

Suppose that after generating feasible solutions from each region $\sigma_1(0)$ is found most promising, then $\sigma(1) = \sigma_1(0)$, and there are two subregions:

$$\sigma_1(1) = \{y \in X : y_{21} = 1, y_{13} = 1\},$$
$$\sigma_2(1) = \{y \in X : y_{21} = 1, y_{13} = 0\},$$

and a surrounding region

$$\sigma_3(1) = \{y \in X : y_{21} = 0\}.$$

This intelligent partitioning then continues in the same manner, going down the ranked list of most important assignments.

### 10.3.2 Generating Feasible Solutions

We know that the ability to quickly generate high-quality feasible solutions from each region is also critical to a successful implementation of the NP method. To address this for the LPDP, we again divide the solution generation into two parts: first we use sampling to address the more difficult decision variables of the problem, and then we use mathematical programming to complete the feasible solution.

Recall that for the LPDP there are two types of decisions: the loads must be assigned to the trucks/drivers and the truck must then be sequenced given the available loads (but does not necessarily need to serve all of the assigned loads). To generate a complete feasible solution both of these decisions must be made. Furthermore, to assure an optimal solution both of them must be made simultaneously. However, within the NP method they can be separated to improve the efficiency while still maintaining the global convergence property.

The following two step algorithm can therefore be used to generate high quality feasible solutions:

**Algorithm** *MP-Sampling*

1. Use random sampling to assign loads to trucks, that is randomly assign values $y_{kl}^0$ to the variables $y_{kl}$ as defined by equation (10.3) above.
2. Given the fixed values $y_{kl}^0$, solve the reduced problem using standard integer programming solver to determine values $s_{lj}^0$ for the remaining sequencing decisions $s_{lj}$.

To illustrate this two part process, we return to the example with two vehicles ($|K| = 2$) and seven loads ($|L| = 7$). To generate a feasible solution (assume no assignment has been fixed by the partitioning), we start by randomly generating an assignment of loads to vehicles. Say, for example, that the randomly generated assignment is $y_{11} = y_{12} = y_{13} = y_{17} = y_{24} = y_{25} = y_{26} = 1$, and $y_{kl} = 0$ otherwise. In other words, Load 1,2,3 and 7 are assigned to Vehicle 1 and Load 4,5 and 6 are assigned to Vehicle 2 (see Figure 10.1).

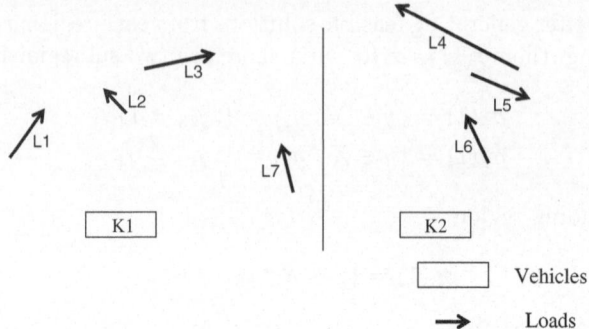

**Fig. 10.1.** Generating a partial solution using random sampling.

To complete this solution an IP is solved for the values of $s_{lj}$ as defined by equation (10.4). In this example, the optimal solution is $s^*_{12} = s^*_{23} = s^*_{65} = s^*_{54} = 1$, and $s^*_{lj} = 0$ otherwise. Thus, the optimal solution is for Vehicle 1 to pickup and deliver Load 1, Load 2, and Load 3 in that sequence, and for Vehicle 2 to pickup and deliver Load 6, Load 5, and Load 4, and that sequence (see Figure 10.2). Load 7 is dropped and no revenue is realized for this load. Note that a complete feasible solution has now been generated, by first sampling to find the values for the load assignments $y$, and then solving an IP to obtain the values of $s$, given these load assignments. The original decision variables $x$ of (10.1) can now be recovered according to $x_{klj} = y_{kl}y_{kj}s_{lj}$, that is,

$$x_{klj} = \begin{cases} 1, \text{ if } k = 1, l = 1, j = 2 \\ 1, \text{ if } k = 1, l = 2, j = 3 \\ 1, \text{ if } k = 2, l = 6, j = 5 \\ 1, \text{ if } k = 2, l = 5, j = 4 \\ 0, \text{ otherwise.} \end{cases} \qquad (10.10)$$

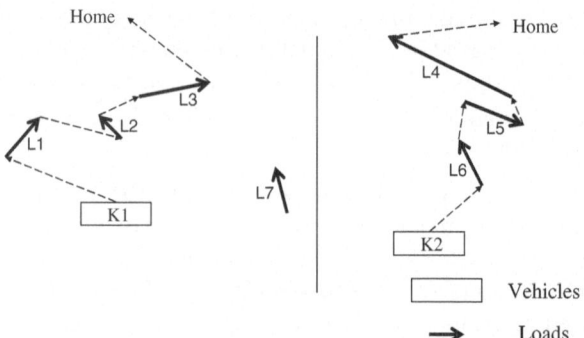

**Fig. 10.2.** Completing the sample solution by solving an IP.

In the first step of the solution generation algorithm, the sampling can be uniform or biased to select more favorable assignments with higher probability. In the second step, since in practical applications relatively few loads are assigned to each truck, solving the sequencing subproblem to optimality can be done quickly using standard software such as CPLEX.

To bias the sampling distribution in the first step we can again use a linear programming relaxation (see Section 4.3.1). Let $(y^{LP}, s^{LP})$ denote the optimal solution found by solving the LP relaxation of the LPDP. Define the weight associated with assigning load $l$ to truck $k$ to be

$$w_{lk} = \frac{y_{kl}^{LP}}{\sum_{h \in K} y_{hl}^{LP}}, \tag{10.11}$$

and then generate load assignment according to this distribution, that is,

$$P[\text{Load } l \text{ is assigned to truck } k] = w_{lk}.$$

We can now put this together in an algorithm for generating high-quality feasible solutions for the LPDP:

**Algorithm** *Biased-MP-Sampling*

1. Use random sampling to assign loads to trucks, that is randomly assign values 0 or 1 to each of the variables $y_{kl}$ according to the following distribution:

$$P[y_{kl} = 1] = w_{lk} = \frac{y_{kl}^{LP}}{\sum_{h \in K} y_{hl}^{LP}}, \tag{10.12}$$

   where as before the solution $y^{LP}$ is found by solving the LP relaxation of the problem.
2. Given the fixed values for the assignment decisions $y$, solve the reduced problem using a standard solver to determine values $s_{lj}^0$ for the remaining sequencing decisions $s$.

In this algorithm, mathematical programming is therefore employed twice: first LP relaxations are used to bias the sampling distribution for assignments, and standard integer programming solver is then used to find the optimal sequence given the assignments. This will assure that high-quality solutions are found quickly.

## 10.4 Numerical Results

Since the LPDP can be formulated as a MIP, we consider using standard solvers to solve it (Wolsey 1998). We have randomly generated test instances and tested them using CPLEX 9.1, which is the state-of-the-art MILP solver, combining standard branch-and-cut, branch-and-bound, and highly efficient meta-heuristics. We examine the performance of the formulations with a time limit of 30 minutes, which is regarded as the typical requirements of the environment this kind of algorithm intended to support. For small and medium size problems with 10–15 trucks, 30–50 loads, and the length of the pickup time window to be Uniform(0,4) hours, CPLEX can provide a result with a satisfactory optimality gap. For large scale problems, CPLEX is not very efficient, typically with an optimality gap greater than 10%.

In this section, we report on our computational experience with the proposed algorithms on randomly generated instances.

### 10.4.1 Test Instances

We randomly generated a set of instances to test the hybrid NP/MP algorithm. The experiment settings are as follows:

- **Map and Locations:** We generate 60 locations in a rectangle map of $X \times Y$ square miles. For each location pair, the length between the two locations is the Euclidean distance on the map.
- **Loads:** Generate loads randomly on the origin-destination location pairs. The handling time of each load is H. The earliest starting time for each load is generated randomly on the time horizon from 7 am to 6 pm, and the length of the pickup time window is set to be Uniform(0,4) hour. The net revenue of serving a load is set to be $V \cdot$(loaded movement time+handling time), where $V$ is the rate of revenue per each service time unit.
- **Vehicles:** For each vehicle, the initial and homing locations are randomly assigned among the locations. (These two locations are not necessarily the same, since our model and algorithm also intent to support some running horizon systems.) Each vehicle's working time is randomly set to be from Uniform(7,9) am to Uniform(4,6) pm individually. The speed of each vehicle is 40 mph. The cost rate of empty movements of vehicles is 10 per time unit.
- **Qualification/Preference:** For each vehicle-load pair $(k, i)$, the probability that vehicle $k$ is qualified and prefer to serve load $i$ is set to $R$.

Overall, we generated 42 test instances, with 6 different groups of parameter setting (as shown in Table 10.1) and 7 different groups of scale setting (as shown in Table 10.2). All these setting are of common properties and scales in real applications. For example, the typical size of local sub fleet handled by a single load manager is around 20, which is the minimum size we use in our test problems.

### 10.4.2 Algorithm Setting

We first solved all of the test instances using CPLEX 9.1 with default CPLEX parameter settings. We implemented our hybrid algorithm in AMPL, and limited the computation time to be within 30 minutes. (For the computation time,

**Table 10.1.** Parameter settings.

| Index | $R$ | $H(h)$ | $(X,Y)$ $(m^2)$ | $V$ |
|-------|-----|--------|-----------------|-----|
| a | 0.8 | 0.1 | (40,40) | Uniform(30,35) |
| b | 0.8 | 0.1 | (40,40) | Uniform(50,60) |
| c | 0.6 | 0.05 | (40,40) | Uniform(30,35) |
| d | 0.8 | 0.05 | (40,40) | Uniform(50,60) |
| e | 0.8 | 0.05 | (60,40) | Uniform(50,60) |
| f | 0.8 | 0.1 | (60,40) | Uniform(50,60) |

**Table 10.2.** Scale setting.

| Index | Number of Trucks | Number of Loads |
|-------|------------------|-----------------|
| 1 | 20 | 60 |
| 2 | 20 | 65 |
| 3 | 25 | 70 |
| 4 | 25 | 75 |
| 5 | 25 | 80 |
| 6 | 30 | 80 |
| 7 | 30 | 90 |

we only calculate the time for the LP solution calculating and partial solution evaluation.) The specifics of the hybrid NP/MP algorithm are described as follow:

- The partitioning is the intelligent partitioning that uses the solution of the LP relaxation to order the vehicle/load assignments.
- For each promising region, the LP relaxation is solved using the dual simplex method.
- Depending on the size of the problem, a total number of 20-100 feasible sample solutions are generated in each iteration.
- Biased random sampling is used to generate vehicle/load assignments for each sample solution. CPLEX is then called to solve the problem associated with that partial solution. Specifically, we apply the value of the current best solution as the feasible bound, set the MIP tolerance gap to be 0.01, and set the computation time limit to be 2 seconds. In our experience, most partial solutions can be completed within one second.
- The algorithm is stopped if one of two criteria is satisfied: (1) The most promising region becomes sufficiently small so that the standard solver can find the optimal solution. (2) If the optimality gap between a known upper bound and the best feasible solution (lower bound) becomes sufficiently small. For many of the test instances, these stopping criteria resulted in the computation time begin much shorter than the 30 minutes time limit.

As a further benchmark for the hybrid NP/MP algorithm, we also implemented and tested a myopic approach that is one most popular method used in applications. This myopic approach is briefly described as follow:

- In each iteration, assign at most one load to each vehicle, and maximize the profits (revenue - empty movement cost) for this stage. Then, let the vehicle serve its assigned load, and update the location and available time of the vehicle. Constraints that guarantee that each can get home on time are added, and assigning a load to a vehicle is only allowed when the profit of the assignment is bigger than a pre-determined parameter $F$. Repeat the above process until no profits can be made.
- Then, $\forall l \in L, k \in K$, if load $l$ is assigned to vehicle $k$, in the original MIP problem, let load $l$ only be available to vehicle $k$ by fixing some variables: $\forall k' \in K.j \in L \cup |L| + |K| + k' : k' \neq k$, fix $x_{k'lj} = 0$. Then resolve the MIP problem to obtain the final schedule.
- For each instance, run the myopic approach for two times with $F$ set to be 0 and 10 respectively, and the better result is selected.

### 10.4.3 Test Results

For 15 of the test instances, it is possible to solve them using CPLEX to an optimality gap less than 10%, that is, if $\overline{z}^{CP}$ is the upper bound from CPLEX, and $\underline{z}^{CP}$ is CPLEX lower bound, that is, the best solution found by CPLEX, the the percentage optimality gap between those upper and lower bounds:

$$z_{GAP}^{CP} = \frac{\overline{z}^{CP} - \underline{z}^{CP}}{\overline{z}^{CP}},$$

satisfies $z_{GAP}^{CP} < 0.1$. We refer to those as the easy instances.

**Table 10.3.** Results for easy instances.

| Ins | $\underline{z}^{CP}$ | $\overline{z}^{CP}$ | $z_{GAP}^{CP}$ | $\underline{z}^{NP}$ | $z_{GAP}^{NP}$ |
|---|---|---|---|---|---|
| a3 | 1236 | 1236 | 0.0 | 1203 | 2.7 |
| a4 | 1265 | 1296 | 2.5 | 1285 | 0.9 |
| b3 | 2137 | 2162 | 1.2 | 2129 | 1.6 |
| c1 | 887 | 899 | 1.4 | 876 | 2.6 |
| c2 | 933 | 987 | 5.8 | 958 | 3.0 |
| c3 | 1045 | 1112 | 6.5 | 1085 | 2.5 |
| d2 | 1728 | 1757 | 1.7 | 1725 | 1.9 |
| d3 | 1795 | 1969 | 9.7 | 1958 | 0.6 |
| e1 | 1911 | 1916 | 0.3 | 1872 | 2.4 |
| e3 | 2166 | 2352 | 8.6 | 2321 | 1.3 |
| e4 | 2428 | 2454 | 1.1 | 2402 | 2.2 |
| f1 | 2076 | 2081 | 0.2 | 2041 | 1.9 |
| f2 | 2272 | 2274 | 0.1 | 2244 | 1.3 |
| f4 | 2653 | 2657 | 0.2 | 2625 | 1.2 |
| f6 | 2753 | 2787 | 1.2 | 2743 | 1.6 |
| Average | | | 2.68 | | 1.85 |

**Table 10.4.** Results for difficult instances.

| Ins | $\underline{z}^{CP}$ | $\overline{z}^{CP}$ | $z_{GAP}^{CP}$ | $\underline{z}^{MY}$ | $z_{GAP}^{MY}$ | $\underline{z}^{NP}$ | $z_{GAP}^{NP}$ | $\Delta z$ |
|-----|------|------|------|------|------|------|------|------|
| a1 | 912 | 1009 | 10.6 | 822 | 22.7 | 992 | 1.7 | 8.8 |
| a2 | 984 | 1102 | 12.0 | 886 | 24.4 | 1067 | 3.3 | 8.4 |
| a5 | 1103 | 1372 | 24.4 | 1089 | 26.0 | 1345 | 2.0 | 21.9 |
| a6 | 1154 | 1361 | 17.9 | 1125 | 21.0 | 1323 | 2.9 | 14.6 |
| a7 | 1169 | 1588 | 35.8 | 1284 | 23.7 | 1549 | 2.5 | 20.6 |
| b1 | 1605 | 1773 | 10.5 | 1529 | 12.1 | 1735 | 2.2 | 8.1 |
| b2 | 1666 | 1935 | 16.1 | 1678 | 15.3 | 1890 | 2.4 | 12.6 |
| b4 | 2004 | 2269 | 13.2 | 1952 | 16.2 | 2240 | 1.3 | 11.8 |
| b5 | 1903 | 2398 | 26.0 | 2123 | 13.0 | 2338 | 2.6 | 10.2 |
| b6 | 2082 | 2374 | 14.0 | 2093 | 13.4 | 2333 | 1.8 | 11.5 |
| b7 | 2153 | 2769 | 28.6 | 2346 | 18.0 | 2725 | 1.6 | 16.2 |
| c4 | 1017 | 1169 | 14.9 | 968 | 20.8 | 1128 | 3.6 | 10.9 |
| c5 | 1029 | 1234 | 20.0 | 1011 | 22.1 | 1186 | 4.0 | 15.3 |
| c6 | 1079 | 1224 | 13.4 | 1039 | 17.8 | 1187 | 3.1 | 10.0 |
| c7 | 1132 | 1432 | 26.5 | 1130 | 26.7 | 1392 | 2.9 | 23.0 |
| d1 | 1445 | 1609 | 11.3 | 1376 | 16.9 | 1568 | 2.6 | 8.5 |
| d4 | 1792 | 2062 | 15.1 | 1774 | 16.2 | 2011 | 2.5 | 12.2 |
| d5 | 1778 | 2178 | 22.5 | 1902 | 14.5 | 2132 | 2.2 | 12.1 |
| d6 | 1802 | 2154 | 19.5 | 1890 | 14.0 | 2129 | 1.2 | 12.6 |
| d7 | 1882 | 2520 | 33.9 | 2191 | 15.0 | 2475 | 1.8 | 13 |
| e2 | 1871 | 2098 | 12.1 | 1918 | 9.4 | 2044 | 2.6 | 6.6 |
| e5 | 2360 | 2607 | 10.4 | 2211 | 17.9 | 2557 | 2.0 | 8.3 |
| e6 | 2055 | 2565 | 24.8 | 2226 | 15.2 | 2509 | 2.2 | 12.7 |
| e7 | 2357 | 3025 | 28.3 | 2576 | 17.8 | 2954 | 2.4 | 14.7 |
| f3 | 2270 | 2541 | 11.9 | 2175 | 16.8 | 2508 | 1.3 | 10.5 |
| f5 | 2387 | 2823 | 18.7 | 2418 | 17.3 | 2743 | 2.9 | 13.4 |
| f7 | 2519 | 3270 | 29.8 | 2786 | 17.4 | 3220 | 1.6 | 15.6 |
| Average | | | 19.46 | | 17.84 | | 2.34 | 12.75 |

We also tested our hybrid approach on these easy instances. The detailed computational results for these 15 instances are shown in Table 10.3. In this table, *Ins* is the instance index, and, similar to the definitions above, $\underline{z}^{NP}$ is the performance of the best solution found by the hybrid NP/MP algorithm, which is thus a lower bound on the optimal performance, and

$$z_{GAP}^{NP} = \frac{\overline{z}^{CP} - \underline{z}^{NP}}{\overline{z}^{CP}}$$

is the percentage optimality gap. The performance of the hybrid NP/MP algorithm on these easy instances is also good, with an average optimality gap of 1.85%. Although the average gap is smaller for the hybrid NP/MP, for some of these instances, CPLEX finds better solutions.

For all other 27 instances, the optimality gap of CPLEX results is greater than 10%, which we consider as difficult instances. We tested our hybrid algorithm and the myopic approach on these instances. The computational results are shown in Table 10.4. In this table, $\underline{z}^{MY}$ is the bound from the best solution of the myopic approach, $z_{GAP}^{MY}$ is the corresponding optimality, and $\Delta z$ is the improvement between the hybrid NP/MP result and the better of the CPLEX and myopic algorithm results, that is,

$$\Delta z = \underline{z}^{NP} - \max\{\underline{z}^{CP}, \underline{z}^{MY}\}.$$

For all these difficult instances, the hybrid NP/MP algorithm outperforms CPLEX and the myopic approach by a significant margin. The solution quality of the hybrid algorithm is very promising, with an average optimality gap of 2.34%, ranging from 1.2–4.0%, compared to an average optimality gap for CPLEX and the myopic algorithm of 19.36% and 17.84%, respectively. Finally, we note that the improvement of our approach is $\overline{\Delta z}$=12.75%.

## 10.5 Conclusions

In this chapter we provide an MIP formulation of the LPDP, which is a difficult problem that arises in logistics operations. We developed an hybrid NP/MP algorithm to solve this problem. This algorithm incorporates mathematical programming into the NP framework in two different ways. First, it uses the solution of an LP relaxation to define an intelligent partitioning method, and second it uses both LP relaxation solutions and the solution to a restricted IP to quickly generate high-quality feasible solutions from each region.

The computational results show that the hybrid NP/MP outperforms applying a state-of-the-art MIP solver directly the problem (CPLEX 9.1). It also performs betters than a myopic heuristic that is frequently used in practice. The results also shows that the advantages of the NP method are greater for difficult problem instances. This is consistent with other results that show that the NP method is most useful for large-scale, complex discrete optimization problems.

# 11

# Extended Job Shop Scheduling

This chapter presents another example of how the NP method can effectively handle realistic problems with very complex constraints. Namely, an extended job shop scheduling problem, where bill-of-material and work-shift constraints are also accounted for in the formulation. To solve this problem, we present an NP algorithm that utilizes intelligent partitioning to impose structure on the search space, and uses an innovative sampling strategy to generate high-quality solutions subject to complex constraints.

## 11.1 Introduction

In a typical shop floor manufacturing environment, one of a scheduler's daily tasks is to assign the released jobs of the day to certain machines with particular performance measures, such as shortest makespan, smallest number of late jobs, lowest cost, or highest throughput. Enterprise Resource Planning (ERP) systems provide a way to easily access all of the necessary information for this task. However, how to efficiently utilize the production resources in a complicated environment still remains a problem for planners and schedulers in factories. Most of the time, the basic form of this assignment process can be modeled as a job shop scheduling problem. The classical job shop scheduling problem has been studied since the 1960's, and is known to be an NP-hard problem. Due to its practical importance, much research has been devoted to the development of efficient solution approaches to tackle this difficult problem. For example, in early work Balas (1969) formulates the job shop scheduling problem in an MIP model without any computational result. Exact methods for solving this MIP are usually based on a branch-and-bound technique. For example, Carlier and Pinson (1989) developed a branch-and-bound method, and for the first time solved the famous 10x10 job shop problem proposed by Fisher and Thompson (1963). Brucker, Jurisch and Sievers (1994) also present a fast branch-and-bound algorithm. Since obtaining exact solutions is often impractical for real problems, many heuristic methods have been proposed to find good solutions instead. One of the most successful approaches is the *shifting bottleneck procedure* of Adams, Balas, and Zawack (1988). This approach improves the schedule by iteratively solving a single *bottleneck* machine problem. Computational results showed that it is effective for many problems. However, implementing or extending the shifting bottleneck procedure is not trivial, which is sometimes a drawback to this approach. Another most applicable and popular approach are various dispatching rules (DR). Blackstone, Philips and Hogg (1982) summarized the main DR available at the time and provided some numerical comparison. Metaheuristics have also been applied extensively to the job shop scheduling problem, and work in this area includes tabu search (Dell'Amico and Trubian 1993, Shi and Pan 2005, Nowicki and Smutnicki 1996) and simulated annealing (Van Laarhoven, Aarts and Lenstra 1992).

As is true for many well-studied problems, a review of the job-shop scheduling literature reveals that the problem often was handled by relaxing many constraints found in the real manufacturing process. The production process in a real shop floor can be much more complicated than the constraints accounted for in the classic job shop formulation, and in particular, we have observed that in addition to the job precedence and machine capacity constraints, there are often constraints on work shifts, bill-of-material, labor availability, multiple machines, and so forth. In this chapter, we address the large-scale extended job shop scheduling problem that adds two of those constraints, namely work shifts and bill-of-material constraints.

## 11.2 Extended Job Shop Formulation

We first consider a job shop scheduling problem with a job set $\mathcal{J}$ and a machine set $\mathcal{M}$. There are $n$ jobs in $\mathcal{J}$. For each job $j \in \mathcal{J}$, the production process of the job follows a predefined operations routine $\mathcal{O}_j = \{1, 2, ..., l_j\}$ (We assume each job $j$ starts from operation 1 and ends at operation $l_j$). Let $\mathcal{O} = \{(j, i) | j \in J, i \in \mathcal{O}_j\}$. Each operation $\mathcal{O}_{ji}$ requires a specified processing time, $p_{ji}$, to run on a particular machine $k$. This is denoted as $\mathcal{O}_{ji}^k$. Let $\mathcal{A}_j = \{(\mathcal{O}_{ji_1}, \mathcal{O}_{ji_2}) | j \in \mathcal{J}, i_1 < i_2\}$ and $\mathcal{E}_k = \{\mathcal{O}_{ji}^k | k \in \mathcal{M}\}$. $\mathcal{A}_j$ represents the precedence relationship between operations in job $j$ and $\mathcal{E}_k$ represents the set of operations running on machine $k$. A schedule of the job shop scheduling problem consists of the operation sequence on each machine and the start and stop time of each operation $\mathcal{O}_{ji}^k$, which are denoted as $S_{ji}$ and $C_{ji}$, receptively. A feasible non-preemptive schedule must satisfy the following conditions:

$$C_{ji} = S_{ji} + p_{ji}, \forall (j, i) \in \mathcal{O} \tag{11.1}$$

$$C_{ji_1} \leq S_{ji_2}, \forall (\mathcal{O}_{ji_1}, \mathcal{O}_{ji_2}) \in \mathcal{A}_j, j \in \mathcal{J} \tag{11.2}$$

$$C_{j_1i_1} \leq S_{j_2i_2} \text{ or } C_{j_2i_2} \leq S_{j_1i_1}, \forall \mathcal{O}_{j_1i_1}, \mathcal{O}_{j_2i_2} \in \mathcal{E}_k, k \in \mathcal{M} \tag{11.3}$$

$$S_{ji} \geq 0, \forall (j, i) \in \mathcal{O} \tag{11.4}$$

In most previous research, the objective of the job shop scheduling problem was to find a non-preemptive schedule with minimal makespan. In our more realistic problem, each job $j$ has a specified due date $d_j$. Our goal is to obtain a non-preemptive schedule with minimal number of late jobs so as to avoid lateness penalties. Let $S_j$ and $C_j$ be the start and stop time of job $j$ (i.e. $S_j = S_{j1}$, $C_j = C_{jl_j}$), respectively. The binary decision variable for job $j$ is defined as

$$U_j = \begin{cases} 1 \text{ if } C_j > d_j \\ 0 \text{ Otherwise} \end{cases}$$

Thus, our objective can be expressed as $\min \sum U_j$.

Some research work in this area was based on the assumption that each job visits a machine no more than once in the scheduling problems. However, we observed that in many production routines, it is not rare that two or more operations of a job are performed on the same machine. These operations can be either adjacent or not. These kinds of jobs are called *reentrant jobs*. Clearly, the reentrant jobs have determined operation orders on the associated machine.

### 11.2.1 Bill-of-Materials Constraints

Sometimes, we may need an operation to assemble different components to build a product. That means some other *subjobs* for the components have to be performed before the main job. This is called *Bill of Materials* (BOM). For example, in order to make a computer, we need to produce many other parts first, such as the motherboard, hard drive, memory, CPU, etc. Then we can assemble these parts to be a complete computer. Figure 11.2.1 shows the relationship between the main job and its subjobs. Job $j_1$ and $j_2$ are the subjobs of job $j$. $l_1$ and $l_2$ are the last operation of $j_1$ and $j_2$, respectively.

The scheduling problem with a BOM was first studied by Balas (1969), who proposed a conceptual MIP model. This model and the algorithm presented in the paper were not tested for large-scale problems due to the lack of adequate computational resources. Czeerwinski and Luh (1994) formulated the BOM constraint by introducing each operation a set which contains all the immediately following operations. Then they used a Lagrangian Relaxation technique to generate a lower bound for the scheduling problem. In our model, we used similar formulation as Czerwinski and Luh (1994). We extended the

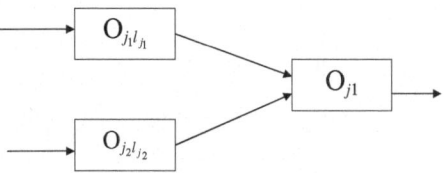

**Fig. 11.1.** Bill of material.

base of set $\mathcal{A}$ from job to operation, because the job precedence constraint is now not a job's only restriction. Specifically, let $\mathcal{A}_{ji}$ be the set of all immediate successors of operation $\mathcal{O}_{ji}$. Accordingly, the constraint (11.2) is changed to:

$$C_{j_1 i_1} \leq S_{j_2 i_2}, \forall \mathcal{O}_{j_2 i_2} \in \mathcal{A}_{j_1 i_1}. \tag{11.5}$$

In practice, there exists one or two levels of subjobs in most manufacturing environments. Hence, the bill of materials constraint changes the data structure of operations in a particular job from *chain* to *intree*. This change complicates the original problem, and consequently, some methods are not applicable with this new setting.

### 11.2.2 Work Shifts Constraints

In the classical job shop scheduling problem, it is always assumed that all of the machines are ready at any time. However, in the real problem, the machine can only work during some specified shifts of a day (usually 8 hours per shift). In addition, a shop floor may be closed during the weekend and the machines in this shop floor therefore stop running during the weekend. For different machines, these working shifts may be different. An operation does not have to be completed during a full working shift, but whenever a machine resumes at the starting time of a shift, it will continue to perform the unfinished operation left on it by the last working shift. This is because of the non-preemptive property of the schedule. An example of working shifts is shown in Figure 11.2.2.

To express the work shifts constraints, we introduce the following parameters and variables.

$T_{mkt}$ = the $k$th working shift for machine $m$. $t$ is binary. $T_{mk0}$ and $T_{mk1}$ are the start time and the stop time of the shift $k$, respectively.
$L_m$ = the number of working shifts for machine $m$.
$X_{jil}$ = the binary variable that specifies the start shift for operation $\mathcal{O}_{ji}$.
$Y_{jil}$ = the binary variable that specifies the stop shift for operation $\mathcal{O}_{ji}$.
$Z_{jil}$ = the binary variable that specifies the working shifts for operation $\mathcal{O}_{ji}$.

Then the formulation can be written as follows:

$$\sum_{l=1}^{L_m} X_{jil} T_{ml0} \leq S_{ji} \leq \sum_{l=1}^{L_m} X_{jil} T_{ml1}, \forall (j,i) \in \mathcal{O}, \mathcal{O}_{ji} \in \mathcal{E}_m \tag{11.6}$$

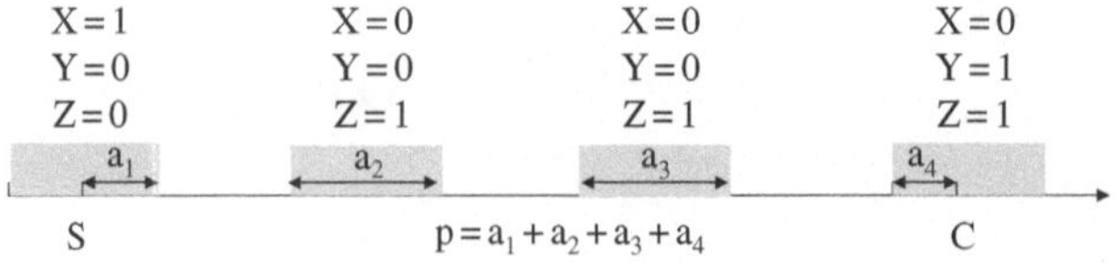

Fig. 11.2. Work shifts.

$$\sum_{l=1}^{L_m} Y_{jil}T_{ml0} \leq C_{ji} \leq \sum_{l=1}^{L_m} Y_{jil}T_{ml1}, \forall(j,i) \in \mathcal{O}, \mathcal{O}_{ji} \in \mathcal{E}_m \qquad (11.7)$$

$$\sum_{l=1}^{L_m} X_{jil} = 1, \forall(j,i) \in \mathcal{O}, \mathcal{O}_{ji} \in \mathcal{E}_m \qquad (11.8)$$

$$\sum_{l=1}^{L_m} Y_{jil} = 1, \forall(j,i) \in \mathcal{O}, \mathcal{O}_{ji} \in \mathcal{E}_m \qquad (11.9)$$

$$\sum_{l=1}^{L_m} X_{jil}T_{ml0} \leq \sum_{l=1}^{L_m} Y_{jil}T_{ml0}, \forall(j,i) \in \mathcal{O}, \mathcal{O}_{ji} \in \mathcal{E}_m \qquad (11.10)$$

$$Z_{ji(l+1)} - Z_{jil} \leq X_{jil}, \forall(j,i) \in \mathcal{O}, l \in [1, L_m - 1], \mathcal{O}_{ji} \in \mathcal{E}_m \qquad (11.11)$$

$$Z_{jil} - Z_{ji(l+1)} \leq Y_{jil}, \forall(j,i) \in \mathcal{O}, l \in [1, L_m - 1], \mathcal{O}_{ji} \in \mathcal{E}_m \qquad (11.12)$$

$$\sum_{l=1}^{L_m} X_{jil}T_{ml1} - S_{ji} +$$

$$\sum_{l=2}^{L_m} Z_{jil}(T_{ml1} - T_{ml0}) \geq p_{ji}, \forall(j,i) \in \mathcal{O}, \mathcal{O}_{ji} \in \mathcal{E}_m \quad (11.13)$$

$$S_{ji} + p_{ji} + \sum_{l=2}^{L_m} Z_{jil}(T_{ml1} - T_{m(l-1)0}) = C_{ji}, \forall(j,i) \in \mathcal{O}, \mathcal{O}_{ji} \in \mathcal{E}_m \ (11.14)$$

Constraint (11.6) and (11.7) specify the start shift and the stop shift of an operation. Constraint (11.8) and (11.9) ensure that an operation can start and stop only once. Constraint (11.10) guarantees the stop shift is located no earlier than the start shift. Constraint (11.11) and (11.12) force the values of $Z$ between the start and stop shifts to be consecutively "1" or "0". Constraint (11.13) allocates enough working shifts for an operation. Constraint (11.14) is a replacement of Constraint (11.1). Assume $X_{jil_1} = 1$ and $Y_{jil_2} = 1$. From the constraints above, we know that $l_1 \leq l_2$ and for any $l \in (l_1, l_2], l_1 < l_2, Z_{jil} = 1$. If $l_1 = l_2$ (i.e. the operation can be completed in one full shift), $Z_{jil_1}$ will be forced to 0.

### 11.2.3 Dispatching Rules (DR)

Considering the job's priority, one might instinctively schedule the job with highest priority first and let it occupy all the required resources, and then schedule the other jobs in the descending order of priority. The priority can be determined by various rules, such as Earliest Due Date (EDD), Shortest Processing Time (SPT), Minimum Slack (MINSLACK), calculated according to $\sigma_j = d_j - C_j$, First Come First Serve (FCFS), Most Work Remaining (MWR). Since our objective is to minimize the number of late jobs, which is due date related, we use the rules of EDD and MINSLACK to sort jobs. In MINSLACK, we estimate the completion time of each job independently. This estimation is computed by accumulating the processing times of the involved operations with the assumption that the associated machines have infinite capacities. Although the operation processing time is dependent on the start time due to the existence of non-working shifts, the completion time is still easily to be estimated since we assume all jobs can start at time 0. Thus, the job that is completed closest to its due date will have the highest priority. Apparently, the solutions generated by DR are always feasible. The DR for our problem can be stated as follows:

**Algorithm** *Dispatching Rules*

1. Sort jobs according to either EDD or MINSLACK. Label the job from 1 to $n$ accordingly. Let $j = 1$.
2. If $j > n$, STOP. Otherwise, let $i = 1$.
3. If $i > l_j$, then $j \leftarrow j + 1$, go to Step 2. Otherwise, on machine $m$, where $O_{ji} \in \mathcal{E}_m$, allocate the earliest free working shifts, which have at least $p_{ji}$ in length, to operation $\mathcal{O}_{ji}$.
4. Let $i \leftarrow i + 1$. Go to Step 3.

Dispatching rules are very quick construction heuristics, but when applied on their own the solution quality is usually less than more elaborate and intelligent heuristics. However, dispatching rules can ge integrated into other techniques to improve the performance of those approaches. For example, Carlier and Pinson (1989) used MWR in their proposed branch-and-bound method for the branching scheme.

## 11.3 NP Method for Extended Job Shop Scheduling

### 11.3.1 Partitioning

As we know, partitioning is the first step in the NP method. It divides the solution space into smaller subregions by fixing some of the decision variables. In the job shop scheduling problem, fixing different operations at a particular position on a machine generates distinct subregions. A sequence of all the positions defines a partitioning scheme. Basically, for a fixed machine order, there are two types of partitioning schemes for the job shop scheduling problem. One that can be termed horizontal and the other that can be termed vertical. Horizontal scheme is based on machine preference. It fixes an operation to each position one by one on machine $k$ first, then considers machine $k + 1$. On the other hand, a vertical scheme is based on position preference. It fixes an operation on position $i$ of all the machines first, then switches to position $i + 1$. In practice, neither one performs better than the other for any instance. But we will later see that the horizontal scheme is more suitable for generating feasible sample solution within the NP method. Thus, we use this scheme in the partitioning step.

We already know that both partitioning schemes are designed on an ordered machine group. But how do we order the machines? Typically, the workload of a given machine plays an important role in scheduling problem. We will have more choices at the very beginning if we partition the most loaded machine first. Therefore, in order to obtain an intelligent partitioning strategy, we first sort all the machines in descending order of the number of distinct jobs (We do not sort machines in the order of the number of operations because of the reentrant jobs). For the distinct operations of the same job on the same machine, we will have to fix the earliest operation that needs to run in the job.

### 11.3.2 Generating Feasible Sample Solutions

For the extended job shop scheduling problem, random sampling is the key to generating feasible sample solutions. In this problem the constraints are very complex, and it is important to keep doing random sampling of feasible points. Sampling of infeasible points in every region will not only obtain useless solutions, but also waste computation resources and eventually reduce the algorithm efficiency. For our problem, the difficulty of sampling feasible points lies in the fact that with prefixed operations on some machines, the orders of some operations on other machines should have been determined. Keeping track of these orders while doing the sampling is apparently not an efficient solution. An algorithm for obtaining a feasible solution without operations prefixed can be found in Pinedo (1995). It develops a solution tree by selecting an operation from a scheduleable operation set $\Omega$ each time. However, this algorithm fails to deal with the situation under partitioning, because it does not track those forbidden operation orders caused by prefixed operations. For example, given machines $M_1$ and $M_2$ and operations $\mathcal{O}_{11}^1$, $\mathcal{O}_{22}^2, \mathcal{O}_{21}^2, \mathcal{O}_{12}^2$, if we prefix operation $\mathcal{O}_{22}^2$, and $\Omega = \{\mathcal{O}_{11}^1, \mathcal{O}_{21}^2\}$, then the selecting and fixing operation $\mathcal{O}_{11}^1$ updates $\Omega = \{\mathcal{O}_{12}^2, \mathcal{O}_{21}^2\}$. If now operation $\mathcal{O}_{12}^2$ is selected to fix, then the solution is infeasible. This shortcoming of the algorithm can be fixed by forbidding the addition of some scheduleable operations to $\Omega$. This is the basic idea of our sampling approach. For the sake of presentation, we introduce the following concepts before the detailed explanation of the approach.

**Definition 11.1.** *A search tree of a machine is an ordered tree with the associated operations as nodes. The root operation is empty. Every child node is an operation that can be run after the operation in the associated parent node is completed.*

**Definition 11.2.** *An operation is free if it has no predecessor or all of its predecessors have already been scheduled.*

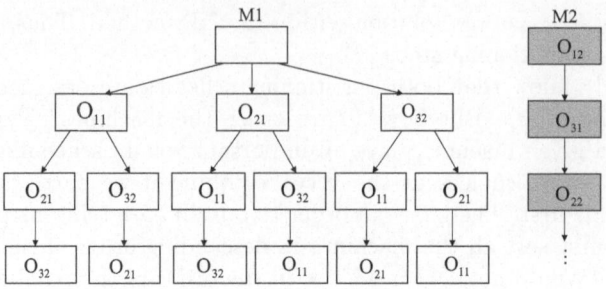

**Fig. 11.3.** Search tree and blocked machine.

**Definition 11.3.** *A machine is blocked if there is a fixed operation that is not free. On a blocked machine j, the first none free fixed operation is called a frozen operation, and its position (starting from 0) is called a frozen point, denoted as* $\mathcal{F}_j$

A search tree enumerates all possible operation sequences on a machine. A path from the root to a leaf represents a feasible job sequence on the machine. Because of the job precedence constraint, some operation orders will possibly become infeasible if an operation on another machine is fixed. Accordingly, some branches of a machine's search tree will possibly be fathomed. Thus, with different operations fixed on machines, the search tree on each machine might be different.

The prefixed operations are also required to be included in the sampling process. In other words, the free prefixed operations will be added into $\Omega$. If a free prefixed operation is selected from $\Omega$, it will be scheduled at the prefixed position. The free operations on a blocked machine are forbidden to be added to $\Omega$, unless they are prefixed and their positions are less than $\mathcal{F}$. When all of the predecessors of a frozen operation have been scheduled, the frozen operation becomes free and the frozen point is moved to the next frozen operation. Meanwhile, all the free operations between the old frozen point and the new one will be added to $\Omega$. The following example shows this process.

Suppose we have two machines $M1$ and $M2$. On $M1$, there are three operations $\mathcal{O}_{11}$, $\mathcal{O}_{21}$, $\mathcal{O}_{32}$. Machine $M2$ is a blocked machine with operations $\mathcal{O}_{12}$, $\mathcal{O}_{31}$ and $\mathcal{O}_{22}$ fixed. Figure 11.3.2 shows the search trees of $M1$ and $M2$. Originally, $\Omega = \{\mathcal{O}_{11}, \mathcal{O}_{21}\}$. $\mathcal{O}_{31}$ is a free operation, but it is not included in $\Omega$ because its position is grater than $\mathcal{F}_2 = 0$. If $\mathcal{O}_{21}$ is selected for schedule, it will not release $\mathcal{O}_{22}$ because $M2$ is blocked with $\mathcal{F}_2 = 0$. If $\mathcal{O}_{11}$ is selected for scheduling, it releases not only operations $\mathcal{O}_{12}$, but also $\mathcal{O}_{31}$ on $M2$. Thus, $\Omega = \{\mathcal{O}_{12}, \mathcal{O}_{31}, \mathcal{O}_{21}\}$. The selection also moves the $M2$'s frozen point to position 2 (i.e. $\mathcal{F}_2 = 2$), where the new frozen operation is $\mathcal{O}_{22}$. By repeating these steps, eventually we will have generated a feasible random solution. The complete approach is presented in the algorithm below.

**Algorithm** *Random-Sampling with Operations Prefixed (RSOP)*

0. Let $P_{ki}$ be the position of the prefixed operation $i$ on machine $k$. For each machine $k$, if all of the prefixed operations are free , $\mathcal{F}_k \leftarrow -1$. Let $\Omega \leftarrow \emptyset$;
1. For each machine $k$, if $\mathcal{F}_k = -1$, include all of the free operations on $k$ to $\Omega$. Otherwise, include every free prefixed operation $i$ where $P_{ki} < \mathcal{F}_k$ to $\Omega$;
2. Randomly select one operation $i \in \Omega$. If $i$ is prefixed, schedule it on the associated position; otherwise, schedule it to the earliest available position on its machine;
3. Let operation $j$ be the immediate successor of $i$ and $j$ is on machine $m$. If $j$ is free, then do the following:
   a) if $\mathcal{F}_m = -1$, add $j$ into $\Omega$. Otherwise
   b) if $P_{mj} = \mathcal{F}_m$ and $G_m$ is the position of the next prefixed operation that is not free, then add all free fixed operations from $\mathcal{F}_m$ to $G_m - 1$ into $\Omega$, $\mathcal{F}_m \leftarrow G_m$; if $G_m$ does not exist, $\mathcal{F}_m \leftarrow -1$, add all left free operations on $m$ into $\Omega$;
4. Let $\Omega \leftarrow \Omega \setminus i$;
5. If $\Omega = \emptyset$, STOP. Otherwise, goto Step 2.

**Proposition 11.4.** *If the prefixed operations define a nonempty solution space, any solution generated by RSOP is feasible.*

*Proof:* First, we prove that the RSOP procedure can always generate a solution (i.e. operation sequence for each machine). Suppose when RSOP finishes (i.e. $\Omega = \emptyset$), there is an unscheduled operation $\mathcal{O}_{ij}$ remained on machine $k$. That implies at least one of its predecessors must be unscheduled on a blocked machine $k_1$ and the frozen operation $\mathcal{O}^{k_1}$ has never been released. Hence, one of $\mathcal{O}^{k_1}$'s predecessors must be unscheduled on another blocked machine $k_2$ with the frozen operation $\mathcal{O}^{k_2}$ unreleased. Repeat this backtracking along the job precedence path; at least one operation $\mathcal{O}^{k_i}$ will be visited twice since the operations set is finite. Then we find a cycle with the job precedence and the precedence forced by the prefixed operations only. Consequently, the associated subregion has an empty solution space, which is a contradiction of our assumption. Second, we prove that every solution generated by RSOP is feasible. Suppose there is a solution that is generated with operations cycle $\mathcal{O}_1 \to \mathcal{O}_2 \to \mathcal{O}_3 \to ... \to \mathcal{O}_1$. Without loss of generality, let operation $\mathcal{O}_1$ be the last scheduled operation. The arc $\mathcal{O}_1 \to \mathcal{O}_2$ does not represent the order performed on the same machine because $\mathcal{O}_2$ is scheduled earlier than $\mathcal{O}_1$. Hence $\mathcal{O}_1$ must be a job predecessor of $\mathcal{O}_2$. This means $\mathcal{O}_2$ is scheduled without its predecessors scheduled first, which violates RSOP procedure. ∎

**Corollary 11.5.** *If RSOP procedure stops with operations unscheduled, the subregion defined by the prefixed operations contains no feasible solution.*

**Proposition 11.6.** *Every feasible solution in the subregion defined by pre-fixed operations has a positive possibility of being generated by RSOP.*

*Proof:* Suppose there is a feasible solution $S$ that cannot be generated by RSOP. Since the subregion is not empty, by Prop. 11.4, RSOP can generate a feasible solution. We follow the RSOP procedure to schedule operations on the exact positions suggested by $S$. This procedure will stop at some stage, in which $\Omega \neq \emptyset$, but there is no operation that can be chosen to fit at the exact position as in $S$. On $S$, let $\Psi = \{\mathcal{O}^k :$ the operation that is immediately after the last scheduled operation on machine $k$ by RSOP$\}$. By assumption, $\Omega \bigcap \Psi = \emptyset$. This implies for any operation $\mathcal{O}^k \in \Psi$, there is an operation $\mathcal{O}^{k'} \in \Psi$ with $k \neq k'$, from which we can start a path to $\mathcal{O}^k$. Because the number of machines is finite, $\Psi$ is finite too. Consequently, there is a cycle that contains two or more operations in $\Psi$. Hence, $S$ is not feasible. This is a contradiction. ∎

**Proposition 11.7.** *The computational complexity of RSOP is $O(|\mathcal{O}|)$, where $|\mathcal{O}|$ is the number of total operations.*

The sampling under uniform distribution is a common method when there is no other information known. However, as for most applications considered in this book, for this problem it is unlikely that every feasible solution is equally possible of being optimal, and is thus likely that a biased random sampling procedure will outperform uniform sampling (see Section 2.3.1). For example, if we sort the operations in the scheduleable operation set $\Omega$ in the ascending order of the jobs' due date, then intuitively, the solution which chooses the first operation from $\Omega$ each time will have a higher probability of being optimal than that which chooses the last operation each time.

The priority of an operation is determined according to the associated job's slack $\sigma_j = d_j - C_j$. The operation with little slack (that is, a tight due date) should potentially be processed earlier. $d_j$ is a constant parameter in our problem and $C_j$ can be estimated by adding the remaining operations' processing time. We already know that the selection of an operation from $\Omega$ may change the search trees of other machines. Thus, the estimated $C_j$ might be changed dynamically. To simplify the procedure, we estimate each $C_j$ at the very beginning (i.e. $C_j = \sum_i p_{ji} + \sum L_{idle}$, where $L_{idle}$ is the length of idle shift between each two adjacent working shifts) and use $\sigma_j$ to guide our weighted sampling. We magnify the $\sigma_j$ with an exponential function so that the operation with minimum slack will be chosen earlier. We also maintain the operations in $\Omega$ in the ascending order of $\sigma_j$ in order to choose the closest due date operation quickly. Define

$$F_{ji} = e^{\alpha/\sigma_j} - 1 \;, \forall \mathcal{O}_{ji} \in \mathcal{O}$$

Theoretically, $\alpha$ can be any positive number. However, if $\alpha$ is too big, $F_{ji}$ will easily overflow on the computer representation. Hence, we let $\alpha \in (0, 1]$ and set a maximum number for $\alpha/\sigma$. Then define

$$P_{ji} = \frac{F_{ji}}{\sum_{\mathcal{O}_{kl} \in \Omega} F_{kl}} , \forall \mathcal{O}_{ji} \in \mathcal{O} \qquad (11.15)$$

To choose a job from $\Omega$, we still need to generate a real number $p \in [0, 1]$. A random number $\mathbf{a}$ can be generated in a uniform distribution $[0, 1]$. To more flexibly adjust the possibility of an operation being selected from $\Omega$, let

$$q = 1 - \mathbf{a}^{\beta} \qquad (11.16)$$

where $\beta \in [0, 1]$. From the ordered set $\Omega$, search the smallest $k$, such that the first $k$ operations' total $P$ is greater than or equal to $q$. Then the $k$th operation is the one to be chosen.

Thus, the algorithm RSOP can be easily changed to WSOP and proposition 11.4 and 11.6 still hold. However, the computational complexity is changed due to the ordered set $\Omega$. The additional steps in WSOP are set $\Omega$ order maintenance (in $O(\log n)$) and operation position locating in $\Omega$ (in $O(n)$). Therefore, we have the following proposition.

**Proposition 11.8.** *The computational complexity of WSOP is $O(|\mathcal{O}|n \log n)$.*

### 11.3.3 Estimating the Promising Index and Backtracking

We already know that the job shop scheduling problem can have different objectives. Given an operation sequence on each machine, we need to assign machine time to each individual operation in order to achieve the predefined goal. This process is called *time tabling*. The strategies of *time tabling* are different for different objectives. For example, the job shop scheduling problem with the objective of minimizing makespan prefers to apply *left justify* schedule, because it advantageous to start each operation as early as possible. But for the problem with the objective of minimizing total inventory, the *right justify* strategy applies. For our objective of minimizing the number of late jobs, it is apparently better to adopt the *left justify* schedule.

As usual, we define the promising index function as the best solution found in the region. The tie is broken arbitrarily if two or more regions are equally promising. If the most promising index region happens to be the surrounding region, we simply backtrack to the root region and start from the beginning. The stop criteria is that either a bottom region is reached or no better solution is found within predefined time.

**Remark 11.3.1** *Since we are considering the problem with working shifts, the duration of each operation does not have to be the fixed processing time. We need to insert machine idle time if necessary.*

Suppose there are $w$ machines in $\mathcal{M}$. Let $\Phi_m$ and $\Gamma_m$ be the sets of all operations and jobs running on machine $m$, respectively. Let $N_m$ be the number of jobs in $\Gamma_m$ and $\Pi_m$ be the fixed operation sequence on machine $m$. Denote the current most promising region as $X$. The NP algorithm for the extended job shop scheduling problem can then be described as below.

**Algorithm** *NP Method*

0. Let $\Pi_m = \emptyset$ for all $m \in \mathcal{M}$. Sort the machines in descending order of $N_m$. Label the sorted machine from 1 to $w$.
1. Let $i = 1$, $X = $ root region.
2. If $i > w$, STOP.
3. Partition $X$ into $N_i$ subregions $X'_1, X'_2, ..., X'_{N_i}$ by fixing the first operation of each job $j \in \Gamma_i$ on machine $i$.
4. Apply RSOP or WSOP on each subregion and the surrounding region.
5. Evaluate each sample and determine the most promising region $X''$. If $X''$ is one of the subregions, then $X = X''$ and goto Step 6. Otherwise, goto Step 7.
6. Suppose $X$ is defined by fixing operation $\mathcal{O}_{jl}$, then add $\mathcal{O}_{jl}$ in $\Pi_i$ and $\Phi_i = \Phi_i \setminus \mathcal{O}_{jl}$. If there is no $\mathcal{O}_{jl'} \in \Phi_i$, then $\Gamma_i = \Gamma_i \setminus j$ and $N_i = N_i - 1$. If $N_i = 0$, then $i = i + 1$. Goto Step 2.
7. Let $\Pi_i = \emptyset$ and reset $\Phi_i$ and $\Gamma_i$ for all $i \in \mathcal{M}$. Goto Step 1.

### 11.3.4 DR-Guided Nested Partitions (NP-DR)

As we will show in Section 11.4, the NP approach is effective for the extended job shop scheduling problem. However, it may be possible to improve it's efficiency further, especially for large-scale problems. One idea to improve the NP efficiency is to reduce the number of sampling points generated in each subregion. But this number cannot be too small, otherwise, the sampled points cannot represent the whole subregion's quality. The improvement of this way is very limited.

Suppose we already know a good solution before executing the NP method. If we are able to completely or partially use this good solution in the NP method, we may obtain better solutions with shorter computation time. Recall that the basic NP method presented in this chapter starts from scratch at the root node, with no operation prefixed, and iteratively searches the subregion until the bottom node is reached. If a good solution is already known before applying NP, then this solution can be used as a guide to prefix some operations in NP (see Section 2.4). In other words, instead of starting from the root node for a new problem, NP can start from a promising subregion, which has a high probability of containing the optimal solution. The guiding solution can be obtained in various ways. For the sake of efficiency, this solution should be achieved quickly. As we describe in Section 11.2.3, dispatching rules are straightforward, and take very little time to obtain a solution. Although the solution obtained by DR is not comparable to that obtained by NP due to the simplicity of the approach most of the times, it is good enough to serve as a guide for our NP approach.

There are many ways to guide NP from a solution. For example, one can fix all the operations that are not late in the solution schedule, and then apply NP to fix those late operations. In our application, we first fix operation sequence on one or more machines according to the guiding solution produced by DR. Then, NP is applied to decide the operation sequences on the other machines. The complete description of NP-DR is as below.

**Algorithm** *NP/DR Hybrid*

0. Obtain a solution $S$ by executing Dispatching Rules;
1. Sort machines in the order of number of operations loaded. Choose the first $k$th machines $(k \leq w)$, fix the operations on these machines in the order suggested by $S$;
2. Starting from the subregion defined from Step 1, apply the algorithm NP[1].

In Algorithm NP/DR Hybrid, if we let $k = 0$, then this is a pure NP method, while if $k = w$, the algorithm turns out to be the DR. We can thus view NP/DR as aa hybrid approach of NP and DR, where DR emphasis algorithm speed and NP aims at good solution quality. Therefore, it is up to the user to trade off between approach speed and solution quality by setting $k$ to an appropriate value.

## 11.4 Computational Results

Our experimental data is from a real industry case. The scheduling length varies from the short term (3 days) to the long term (60 days). Based on the number of jobs involved, we classify the instances in 3 groups: small-scale $(n \leq 100)$, mid-scale$(100 < n \leq 400)$ and large-scale $(400 < n \leq 800)$. For each group, we select 5 instances to test. The details of the instances we tested are shown in Table 11.1. Usually, a job contains 4 to 5 operations, but because some of the operations are virtual operations (e.g. shipping), which do not really consume factory resources, we ignore these operations. Hence, in a typical job shop scheduling problem, one job has 2 to 3 operations on average. The "$|\mathcal{O}|_{sub}$" in Table 11.1 shows the number of operations in the jobs to finish the bill of materials. Although we treat these jobs the same as the other jobs from customers, we actually do not have specified due dates for them. In order to be consistent with the problem setting, we manually assign due dates to these jobs (usually 7 days ahead of the due date of the supporting jobs). But any delay of these jobs is not counted in our final schedule.

---

[1] In case of backtracking, the approach does not return back to the root node, but to the subregion that the NP-DR starts from.

**Table 11.1.** Instances setting.

| | Small-Scale | | | | |
|---|---|---|---|---|---|
| Instance | 1 | 2 | 3 | 4 | 5 |
| $n$ | 28 | 30 | 91 | 72 | 52 |
| $|\mathcal{O}|$ | 64 | 54 | 140 | 132 | 104 |
| $|\mathcal{O}|_{sub}$ | 2 | 16 | 11 | 11 | 10 |
| $w$ | 27 | 19 | 43 | 47 | 44 |
| | Mid-Scale | | | | |
| Instance | 6 | 7 | 8 | 9 | 10 |
| $n$ | 224 | 225 | 266 | 204 | 361 |
| $|\mathcal{O}|$ | 459 | 532 | 480 | 443 | 801 |
| $|\mathcal{O}|_{sub}$ | 65 | 60 | 46 | 61 | 68 |
| $w$ | 97 | 90 | 89 | 87 | 101 |
| | Large-Scale | | | | |
| Instance | 11 | 12 | 13 | 14 | 15 |
| $n$ | 423 | 473 | 606 | 664 | 725 |
| $|\mathcal{O}|$ | 906 | 1028 | 1220 | 1440 | 1601 |
| $|\mathcal{O}|_{sub}$ | 94 | 107 | 113 | 131 | 142 |
| $w$ | 115 | 117 | 128 | 132 | 135 |

**Table 11.2.** Problem size.

| Instance | Number of Variables | Number of Constraints |
|---|---|---|
| 1 | 15382 | 10900 |
| 2 | 16231 | 21334 |
| 3 | 33662 | 24393 |
| 4 | 31888 | 21870 |
| 5 | 24397 | 17371 |
| 6 | 508937 | 347428 |
| 7 | 764565 | 522257 |
| 8 | 386568 | 271341 |
| 9 | 366822 | 259137 |
| 10 | 1455030 | 1008630 |
| 11 | 1499550 | 1031470 |
| 12 | 1221940 | 851892 |
| 13 | 1049080 | 762969 |
| 14 | 2416490 | 1689110 |
| 15 | 2034000 | 1456200 |

We coded our algorithm in C++. All the instances were tested on a Pentium IV 3.2GHz CPU. The numerical results of the two dispatching rules EDD and MINSLACK are shown in Table 11.3. We demonstrated the solutions by dispatching rules EDD and MINSLACK, which were obtained in almost no time. We also give the lower bounds for parts of the instances. These lower bounds were calculated by solving the MIP model in Section 11.2. We modeled the extended job shop scheduling problem in AMPL and solved by CPLEX.

**Table 11.3.** Dispatching rules solutions and lower bounds.

| Ins | EDD $\sum U_j$ | CPU(sec) | Gap (%) | MINSLACK $\sum U_j$ | CPU | Gap | Lower Bound |
|-----|-----|-----|-----|-----|-----|-----|-----|
| 1 | **1** | 1 | 0 | **1** | 1 | 0 | 1 |
| 2 | 3 | 1 | 50 | 3 | 1 | 50 | 2 |
| 3 | 28 | 1 | 56 | 28 | 1 | 56 | 18 |
| 4 | 14 | 1 | 133 | 15 | 1 | 150 | 6 |
| 5 | 8 | 1 | 167 | 6 | 1 | 100 | 3 |
| 6 | 47 | 1 | 135 | 43 | 1 | 115 | 20 |
| 7 | 75 | 1 | 150 | 70 | 1 | 129 | 39 |
| 8 | 19 | 1 | 533 | 18 | 1 | 500 | 3 |
| 9 | 62 | 1 | 313 | 59 | 1 | 293 | 15 |
| 10 | 102 | 1 | - | 100 | 1 | - | - |
| 11 | 104 | 1 | - | 102 | 1 | - | - |
| 12 | 106 | 1 | - | 106 | 1 | - | - |
| 13 | 182 | 2 | - | 163 | 2 | - | - |
| 14 | 194 | 2 | - | 188 | 2 | - | - |
| 15 | 201 | 2 | - | 193 | 2 | - | - |

Number in bold is the optimal value.

However, due to the size of the problem (see Table 11.2), none of the testing instances can be solved completely by applying this model directly. Only by relaxing parts or all of the binary variables $(X, Y, Z)$ for time slots were we able to obtain the lower bounds for some small-scale instances. All the lower bounds shown were the result of executing the MIP model for 2 hours.

### 11.4.1 Effectiveness of Weighted Sampling

Table 11.4 shows the results of NP-WSOP with different $\alpha$ and $\beta$. Compared to the results in Table 11.3, NP-WSOP is much better than EDD and MINSLACK in most instances. Some of the solutions can even reach optimality. With the guide of $\sigma$, NP-WSOP actually concentrates its limited samples on the solutions with the highest probability of being optimal. In fact, many effective NP applications are embed weighted sampling. Hence, for our problem, we prefer to use NP-WSOP rather than NP-RSOP. Keep in mind that it does not mean RSOP is meaningless for this NP application. On the contrary, the RSOP provides a way to detect the infeasible region (Cor. 11.5) and a way to sample the feasible solutions in a subregion (Prop. 11.4 and 11.6). WSOP improves upon RSOP by introducing $\sigma$, with the purpose of locating more accurate subregion within a small number of samples.

Both NP-WSOP and the dispatching rules EDD and MINSLACK take advantage of external information. Because NP-WSOP is intelligent in searching in each solution space and samples more solutions than the dispatching rules, it is not surprising that it outperforms EDD and MINSLACK on most of the problems. The cost of this improvement is computation time. This difficulty

**Table 11.4.** Solutions found by the NP method.

| Ins | $\alpha = 1, \beta = 0.125$ | | $\alpha = 0.6, \beta = 0.25$ | | $\alpha = 0.25, \beta = 0.8$ | | $\beta = 0$ | |
|---|---|---|---|---|---|---|---|---|
| | $\sum U_j$ | CPU(sec) | $\sum U_j$ | CPU | $\sum U_j$ | CPU | $\sum U_j$ | CPU |
| 1 | **1** | 1 | **1** | 1 | **1** | 1 | **1** | 1 |
| 2 | 3 | 1 | 3 | 1 | 3 | 1 | 3 | 1 |
| 3 | 29 | 1 | 33 | 2 | 22 | 2 | 24 | 1 |
| 4 | 19 | 1 | 19 | 1 | 9 | 1 | 14 | 1 |
| 5 | 6 | 1 | 6 | 1 | 6 | 1 | 5 | 1 |
| 6 | 37 | 18 | 35 | 18 | 35 | 18 | 33 | 7 |
| 7 | 80 | 27 | 102 | 28 | 85 | 28 | 86 | 10 |
| 8 | **3** | 245 | **3** | 229 | **3** | 237 | **3** | 74 |
| 9 | 83 | 29 | 85 | 30 | 83 | 32 | 66 | 6 |
| 10 | 107 | 34 | 109 | 35 | 104 | 37 | 90 | 7 |
| 11 | 72 | 174 | 77 | 178 | 85 | 187 | 64 | 58 |
| 12 | 114 | 222 | 94 | 228 | 93 | 242 | 112 | 74 |
| 13 | 127 | 576 | 126 | 592 | 147 | 637 | 111 | 95 |
| 14 | 158 | 828 | 177 | 849 | 171 | 907 | 135 | 272 |
| 15 | 200 | 1200 | 198 | 1222 | 195 | 1365 | 176 | 395 |

can be remedied by applying NP-DR. As shown in Table 11.5, by fixing operations in a number of machines according to the solutions produced by DR, NP-DR quickly determines its starting subregion, and dramatically decreases the computation time. The solutions of NP-DR is guaranteed to be no worse than DR. Hence, in some special cases, such as the 7th instance, when DR outperforms NP-WSOP, NP-DR obtains better solutions based on the ones produced by DR.

### 11.4.2 $\alpha$ Sensitivity

The exponential function parameter $\alpha$ is introduced to magnify the possibility that the operation with the tightest due date will be selected from $\Omega$. It can be seen that $P_{ji}$ for the operation with the smallest $\sigma$ increases dramatically as $\alpha$ increases, while on the other hand, all the other $P_{ji}$ decrease quickly. Theoretically, if $\alpha \to +\infty$, we actually choose the operation with smallest $\sigma_j$ every time from $\Omega$. In this situation, our NP method needs to check only one sample in each subregion. As a result, it solves mid and large scale problems much more quickly than the normal NP approach.

Generally, NP-WOSP with $\alpha \to +\infty$ can find good solutions. However, not all optimal solutions require the most urgent operation to be scheduled first. For many scheduling problems, giving up the most urgent job leaves more available resources for others, which may help more of the other jobs to be completed on time. Moreover, recall that we use a static $\sigma$ instead of a dynamic $\sigma$. Considering the error between the estimated and the real most

**Table 11.5.** Solutions found by the Hybrid NP/DR method.

| Ins | NP-DR($\alpha = 0.6, \beta = 0.25$) | | | Best NP Solution | |
|---|---|---|---|---|---|
| | $\sum U_j$ | Fixed Machines | CPU(sec) | Gap (%) | Improved(%)* |
| 1 | 1 | 2 | 1 | 0 | 0 |
| 2 | 3 | 2 | 1 | 50 | 0 |
| 3 | 20 | 2 | 1 | 11 | 44 |
| 4 | 15 | 2 | 1 | 50 | 83 |
| 5 | 5 | 2 | 1 | 67 | 100 |
| 6 | 32 | 2 | 13 | 60 | 55 |
| 7 | 69 | 8 | 7 | 125 | 4 |
| 8 | 5 | 3 | 77 | 0 | 500 |
| 9 | 57 | 13 | 3 | 280 | 33 |
| 10 | 87 | 12 | 5 | - | 13 |
| 11 | 93 | 3 | 106 | - | 37 |
| 12 | 106 | 3 | 148 | - | 12 |
| 13 | 107 | 20 | 33 | - | 34 |
| 14 | 184 | 5 | 355 | - | 28 |
| 15 | 193 | 10 | 218 | - | 8 |

*Compared with DR. For Instance 1 to 9, the improved is the difference of gap; from 10 to 15, the improved is calculated by $100 * (\min_{DR} \sum U_j - \min_{NP} \sum U_j)/\min_{DR} \sum U_j$.

urgent jobs, the static $\sigma$ does not guarantee that we choose the exact job from $\Omega$ every time.

Figure 11.4.3 demonstrates how $\alpha$ affects the number of late jobs with fixed $\beta$ for the 12th instance. This figure clearly shows that the best solution is not found at the largest $\alpha$ (in our experiment setting, $\alpha \in (0, 1]$). The best $\alpha$ is different for different problems.

### 11.4.3 $\beta$ Sensitivity

If the random variable $q$ is uniformly selected in $[0,1]$, then $P_{ji}$ defined in Eq. (11.15) exactly represents the probability that operation $\mathcal{O}_{ji}$ can be chosen. As stated in Section 11.4.2, the value of $P_{ji}$ can be adjusted by altering the value of $\alpha$. However, we already indicate in Section 11.3.2 that $\alpha$ cannot be too big because of the capacity of the computer representation. Then in some cases in which $\alpha$ is not big enough to ensure the high possibility that the operation with the tightest due date will be chosen, it is necessary to adjust $\beta$ as a supplement. The smaller the $\beta$, the higher the probability that the operation with tightest due date will be chosen.

In fact, in Eq. (11.16), if $\beta = 1$, we have regular random values in uniform distribution; but if $\beta = 0$, then $q \equiv 0$, which means we always choose the first operation in $\Omega$. This is equivalent to letting $\alpha \to +\infty$. We have already discussed this situation in Section 11.4.2. Figure 11.4.3 shows the different results with different $\beta$ for the 12th instance. From this figure, we can see

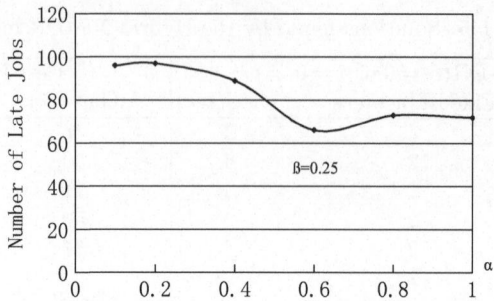

**Fig. 11.4.** $\alpha$ Sensitivity for the 12th instance.

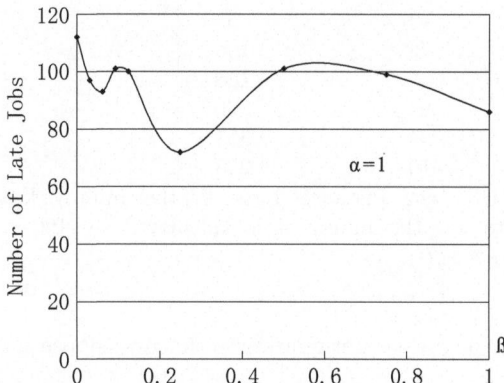

**Fig. 11.5.** $\beta$ Sensitivity for the 12th instance.

that the solution at either end is not the best. Similar to the selection of $\alpha$, for most instances, the best $\beta$ is found between 0 and 1.

It looks like redundant to use two parameters to express a probability function in the weighted sampling. However, we introduce $\alpha$ and $\beta$ for different purposes. We have already explained the reason of applying $\beta$. One might ask if we can use $\beta$ only and ignore $\alpha$. Remember that the purpose of the exponential function with parameter $\alpha$ is to magnify the possibility that the operation with the tightest due date will be present in the ordered set $\Omega$. Apparently, $\beta$ does not have such a feature. Therefore, we include both $\alpha$ and $\beta$ together in our NP algorithm with WSOP.

## 11.5 Conclusions

This chapter illustrates the NP method for a very complex real application, namely the classic job shop scheduling problem with additional constraints that account for realistic requirements usually absent in job shop formulations. As has been done elsewhere in this book, the keys to a successful implementation is to (a) design a method for intelligent partitioning that imposes a structure on the feasible region, and (b) design an efficient method for generating high-quality feasible sample solutions. Both are achieved in this chapter.

The resulting NP algorithm was tested on real industry date. The results show that all of the tested problems can be solved by NP efficiently. Note that the time for solving the largest scale instance is less than 30 minutes. This computation time is acceptable by most industrial schedulers, especially those in median sized businesses, thus illustrating the NP algorithms capabilities for solving real industry problems within acceptable time.

# Resource Allocation under Uncertainty

## 12.1 Introduction

In this final chapter of the book, we consider a resource allocation problem where the objective function is inherently noisy. Resource allocation problems is a very general class of problem, and may include problems such as facility planning, job scheduling, buffer allocation, pollution control, and portfolio management. Many such problems fall into the applicable areas of stochastic discrete optimization. Owing to the complexity inherent in these systems, the search for optimal solutions can be daunting. Two of the key difficulties for solving the problem are: (1) the combinatorial explosion of alternatives normally leads to NP-hard optimization problems; (2) the lack of analytical expressions relating performance functions to solutions usually results in noisy estimates of the performances. The first of these has been extensively explored in this book, and as outlined in Chapter 3, the NP method is also applicable when the objective function is noisy.

## 12.2 Optimal Computing Budget Allocation

In Section 3.2.1, we introduced the concept of ordinal optimization and the connections between the NP method and ordinal optimization (Ho, Sreenivas and Vakili 1992). In this chapter, we will build further on those concepts and combine them with an efficient simulation control technique (Chen et al. 1997). Recall that from the perspective of ordinal optimization, the NP method uses order comparison to determine the most promising region from a set of sample solution that it generates. The benefits of only needing ordinal comparisons stem from the fact that it has been shown (Dai 1996) that the convergence rate of the ordinal comparison can be exponential as the number of simulation replications or samples increases while the convergence rate of the cardinal value estimate is at most $O(1/\sqrt{t})$, where $t$ is the number of simulation replications for a terminating simulation, or the number of simulation samples for a steady-state simulation (Fabian 1971, Kushner and Clark 1978) This characteristic lends itself readily to simulation-based ordinal comparison approach.

While *ordinal optimization* could significantly reduce the computational cost for discrete event systems simulation, there is potential for further improvement of its performance by intelligently determining the numbers of simulation samples (or replications) among different solutions. Intuitively, to ensure a high alignment probability, a large portion of the computing budget should be allocated to those solutions that are critical in the process of identifying superior solutions. In other words, a large number of simulations must be conducted with those critical solutions in order to reduce variance of ordinal comparison. Conversely, efforts ought to be minimized on computations devoted to non-critical solutions exerting little or no effect on identifying superior solutions even though these solutions boast large variances. In so doing, the overall simulation efficiency is improved. Ideally, we want to optimally choose the number of simulation samples (or replications) for all solutions to maximize simulation efficiency with a given computing budget. OCBA (*optimal computing budget allocation*) is one of such techniques that can be used in a simulation-optimization procedure for efficiently ranking and selecting the most appropriate solution (Chen et al. 1996).

As we can see, the NP method focuses on selection of solutions, whereas ordinal optimization and OCBA emphasize comparison of selected solutions. Functionally, these two approaches are mutually complimentary. We apply this hybrid algorithm for a stochastic resource allocation problem where no analytical expression exists for the objective function and where the problem can only be estimated through simulation. Numerical results show that our proposed algorithm can be effectively used for solving large-scale stochastic discrete optimization problems.

## 12.3 Stochastic Resource Allocation Problems

There are many resource allocation problems in the design of discrete event systems. The following examples are two typical cases in point.

**Example 1. Buffer Allocation Problem in Communication Networks**
We consider a 10-node network shown in Figure 12.1 (details can be found in Chen and Ho 1995). There are 10 servers and 10 buffers that are interconnected in a switching network. We then assume that there are two classes of customers with different arrival distributions, but the same service requirements. We consider both exponential and non-exponential distributions (uniform) in the network. Both classes arrive at any of Nodes 0-3, and leave the network after having gone through three different stages of service. Rather than probabilistic, the routing is class dependent as shown in Figure 12.1.

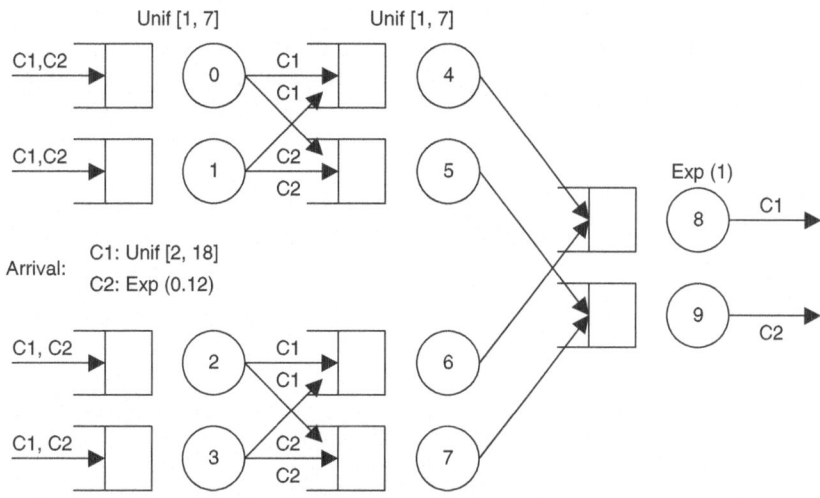

**Fig. 12.1.** A 10-node network.

Finite buffer sizes at all nodes are assumed which is exactly what makes our optimization problem interesting. Specifically, we are interested in distributing optimally a given number of buffer spaces to different nodes so that the network throughput is maximized. A buffer is said to be full if and when there are as many customers as its capacity allows, not including the customer being tended to in the server. We consider the problem of allocating 12 buffer units, among the 10 different nodes numbered from 0 to 9. We denote the buffer size of node $i$ by $x_i$. Specifically,

$$x_0 + x_1 + x_2 + \ldots + x_9 = 12, \text{ and } x_i \text{ is a non-negative integer} \qquad (12.1)$$

The number of different combinations of $[x_0, x_1, x_2, \ldots, x_9]$ which satisfy the constraint in (12.1) can be calculated as follows:

$$\binom{12 + 10 - 1}{10 - 1} = 293,930.$$

Unfortunately, due to the dynamic nature of the system, there is no closed-form analytical formula to evaluate the performance function (throughput, in this example). For each combination, the performance measure estimation involves a very long simulation (for steady state simulation) or a huge number of independent replications (for transient simulation). The total simulation cost is prohibitively high even if the simulation cost for a single solution alternative is low. In Section 12.5, we will illustrate the benefits of using the proposed algorithm to this buffer allocation problem.

**Example 2. Resource Allocation in Manufacturing Systems**

A manufacturing system consists of $C$ manufacturing cells and a total of $R$ resources that can be allocated to any of the manufacturing cells. What we seek is the optimal allocation of resources to manufacturing cells. Given a performance function (or measure) $J(\cdot)$, we can formulate this problem as follows:

$$\max_{x \in X} J(x = x_1, x_2, \ldots, x_C) \tag{12.2}$$

s.t.

$$x_1 + x_2 + \ldots + x_C \leq R \text{ and } \forall x_i \geq 1.$$

As can be seen, the solution alternatives grow exponentially when the problem size increases. For example, if we let $C = 10$ and $R = 30$, then the total number of different combinations (solutions) is $\sum_{m=C}^{R} \binom{m-1}{C-1} = 30,045,015$.

Tackling the type of resource allocation problems discussed above is the main focus of this paper. It should be noted that we do not assume any special property with respect to the performance function. In addition, due to the inherent complexity of resource allocation problems, we expect that most performance functions cannot be defined by an analytical expression. Often $J(x)$ is an expectation of some random estimate of the performance,

$$J(x) = E[L(x, \xi)] \tag{12.3}$$

where $\xi$ is a random vector that represents uncertain factors in the systems, $x \in X$ and $X$ is a discrete and finite set. To be able to estimate $J(x)$, two main approaches are available: analytic approximation and simulation. In this paper, we consider using discrete event simulation, i.e., to estimate $E[L(x, \xi)]$ by

$$E[L(x, \xi)] \approx \hat{J}(x) \equiv \frac{1}{t} \sum_{i=1}^{t} L(x, \xi_i). \tag{12.4}$$

Unfortunately, $t$ cannot be too small for a reasonable estimation of $E[L(x, \xi)]$. Thus, obtaining an optimal solution for resource allocation problems becomes an arduous task when the solution space is presumably very large. In the next section, we will present an effective approach for this type of resource allocation problems.

## 12.4 NP Method for Resource Allocation

It is well known that simulation could be very time consuming when it comes to a complicated objective function. In particular, as the number of alternative solutions grows, so does the total simulation cost. To remedy the situation,

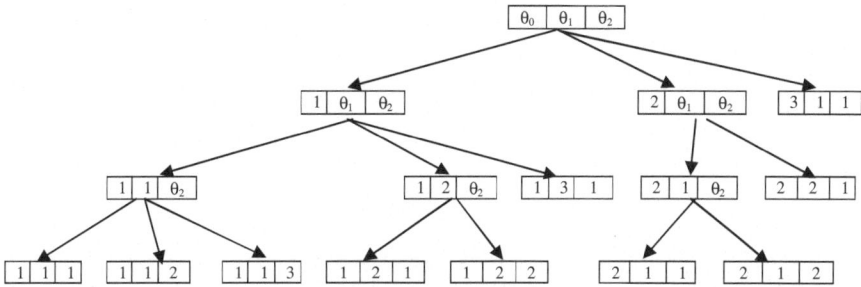

**Fig. 12.2.** Partitioning of the manufacturing resource allocation problem.

our proposed hybrid approach integrates the NP method, ordinal comparison, and OCBA for ranking and selection. Among the merits of this approach, ordinal comparison has emerged as an efficient technique for simulation and optimization, converging exponentially in many cases. The OCBA further enhances the efficiency of ordinal optimization by intelligently determining the best allocation of simulation replications or samples necessary to maximize the probability of identifying the optimal ordinal solution. The integration of these methods within a hybrid NP method enables the combined approach to be applied to large-scale discrete optimization problems.

As for any other NP hybrid, the first step is to partition the current most promising region. This partitioning strategy imposes a structure on the feasible region and is therefore very important for the rate of convergence of the algorithm. As an illustration, we can partition the solution space of Example 2 as showed in Figure 12.2, given that $R = 5$ and $C = 3$. It should be noted that the partitioning showed here is completely unrelated to the performance function, so this is a generic partitioning approach. As we have encountered for other applications, more efficient partitions could be constructed if the performance function is considered, that is, by developing intelligent partitioning (see Section 2.2). This will be done later.

The other key element to a successful NP implementation is a method for efficiently generating high-quality sample solutions. Although we know that uniform sampling can always be used, we also know that it is usually worthwhile to incorporate special structures into the sampling procedure so that the solution quality can be improved. For example, we can use a weighted sampling scheme to generate a sample point for the resource allocation problem discussed in Example 1. The weight can be determined by carefully considering the system structure so that some solutions have a higher probability of being chosen than other solutions. Such a biased sampling scheme will be discussed below.

### 12.4.1 Calculating the Promising Index through Ordinal Optimization

After generating feasible sample solution from each region, the next step of the NP algorithm is to calculate the promising index of each region. As we saw in

Chapter 2 and throughout the application chapters of the book, a commonly used promising index is

$$I(\sigma) \equiv \max_{x \in \sigma} J(x). \tag{12.5}$$

where $\sigma$ is a subregion of $X$. To estimate the promising index, we let

$$\hat{I}(\sigma) \equiv \max_{x \in \{\text{the sampled solutions in } \sigma\}} \hat{J}(x). \tag{12.6}$$

It follows that the region with that best sample solution becomes the new promising region in the next iteration. Let $H(k)$ be the set which collects all sample solutions to be simulated in the $k$-th iteration. Namely $H(k)$ is the union of all sample solutions from all subregions. Even though the size of $H(k)$ is much smaller than the entire solution space, it is still very time consuming to evaluate all the solutions in $H(k)$. For example, if we have 10 subregions in each iteration and we have 15 sample solutions for each subregion, then we have to simulate 150 solutions totally at each iteration in order to determine the most promising region.

With traditional simulation methods, the accuracy, as measure by the width of a confidence interval around the estimate $\hat{J}(x)$, cannot improve faster than $O(1/\sqrt{t})$, the result of averaging i.i.d. noise. To obtain an acceptable statistical estimate for a solution, a large $t$ is therefore usually required for each solution, again implying a long computation time.

Although the variance of the estimate of $\hat{J}(x)$ decays slowly as $t$ goes to infinity, recent research has shown that comparing *relative orders* of performance measures converges much faster than the performance measures themselves do. As stated above, this is the basic idea of *ordinal optimization* (Ho et al. 1992). Dai (1996) shows that under certain conditions the rate of convergence for ordinal comparison can be exponential (details are given in Theorem 12.1). A significant implication of this observed result is that a good estimate on the relative order of compared solutions can be obtained when the value estimate remains poor. This idea is applicable not only to problems with discrete solution space, but also to problems over a continuous solution space (Cassandras and Bao 1994, Cassandras and Julka 1994, Chen et al. 1998, Chen et al. 1999, Gong et al. 1995, Patsis et al. 1997, Yan and Mukai 1993).

As we can see from (12.6) and we have noted before, the NP method is based on order comparison. At each NP iteration, we select the region that contains the best solution $x^{a(k)}$ using the following criterion:

$$x^{a(k)} \equiv \arg \max_{x \in H(k)} \hat{J}(x) (= \frac{1}{t} \sum_{i=1}^{t} L(x, \xi_i)). \tag{12.7}$$

For notational simplicity, we use $x^a$ a rather than $x^{a(k)}$ in the remainder of the chapter. Given the fact that we use only a finite number of simulation replications, $t$, $\hat{J}(x)$ is an approximation to the true expected performance $E[L(x, \xi)]$. The solution $x^a$ with the largest value of $\hat{J}(x)$ is not necessarily the true best solution. Thus, we introduce the following concept.

**Definition 12.1.** *Define correct selection (CS) as the event that the selected solution $x^a$ is actually the best solution in $H(k)$. Define the correct selection probability $P(CS) \equiv P\{$ The current top-ranking solution $x^a$ is actually the best among the simulated solutions$\}$.*

Based on the results from *ordinal comparison* (Dai 1996), it is possible to establish *relative order* of $\hat{J}(x)$ efficiently (i.e., to make the probability $P(CS)$ sufficiently high) although the variance of $\hat{J}(x)$ may attrition slowly.

**Theorem 12.2.** *Suppose the simulation samples for each solution are i.i.d. and the simulation samples between any two solutions are independent. Assume that $L(x, \xi_i)$ has a finite moment generating function. The ordinal comparison confidence probability converges to 1 exponentially. More specifically, there are $\alpha > 0$, $\beta > 0$ such that*

$$P(CS) \geq 1 - \alpha e^{-\beta t}$$

*Proof.* See Theorem 5.1 in Dai (1996).

Since most statistical distributions (for example, normal, exponential, Erlang, and uniform distributions) have finite moment generating functions, Theorem 12.1 is valid in most cases. However, it is not easy to estimate $\alpha$ and $\beta$, which are functions of the relative differences among all solutions and their variances. While it is possible to estimate $P(CS)$ using an extra Monte Carlo simulation, it is too expensive given this setting. Under a Bayesian model, Chen (1996) develops an estimation technique to quantify the confidence level for ordinal comparison $P(CS)$, which is presented as follows:

**Theorem 12.3.** *Let $\tilde{J}_x$ denote the random variable whose probability distribution is the posterior distribution of the expected performance for solution $x$ under a Bayesian model. Assume $\tilde{J}_{x^i}$ and $\tilde{J}_{x^j}$, $i \neq j$, are independent (i.e., the simulations for different solutions are independent). For a maximization problem,*

$$P(CS) \approx \prod_{j \neq a} P\{\tilde{J}_{x^a} > \tilde{J}_{x^j}\}$$

$\equiv$ Approximate Probability of Correct Selection (APCS).

Under the Bayesian model, the posterior distribution $p(\tilde{J}_x)$ consists of information from both the prior distribution and the simulation results $\{L(x, \xi_i), i = 1, 2, \ldots, t\}$. Furthermore, with a mild Gaussian assumption, if the variance $\sigma_x^2$ is known (Bernardo and Smith 1995),

$$\tilde{J}_x \sim N\left(\frac{1}{t_x} \sum_{i=1}^{t_x} L(x, \xi_i), \frac{\sigma_x^2}{t_x}\right)$$

Then,

$$APCS = \prod_{j \neq a} P\{\tilde{J}_{x^a} > \tilde{J}_{x^j}\}$$

$$= \prod_{j \neq a} \Phi \left( \frac{\frac{1}{t_a}\sum_{i=1}^{t_a} L(x^a, \xi_i) - \frac{1}{t_j}\sum_{i=1}^{t_j} L(x^j, \xi_i)}{\sqrt{\frac{\sigma_a^2}{t_a} + \frac{\sigma_j^2}{t_j}}} \right). \qquad (12.8)$$

where $\Phi$ is the standard normal cumulative distribution function and $t_x$, $t_a$, $t_i$, and $t_j$ are simulation numbers. The $APCS$ in (12.8) gives an estimate of the probability that the selected best solution is indeed the true best solution. Note that the computation of $APCS$ is simply a product of pairwise comparison probabilities. If the variance is unknown, $\sigma_x^2$ can be replaced by the sample variance and $\tilde{J}_x$ becomes $t$-distributed (Chick 1997). Empirical testing shows that $APCS$ provides a good approximation for $P(CS)$ (Chen 1996, Chen et al. 1998b, 2000a, Inoue and Chick 1998, Chick et al. 1999). For this reason, we adopt $APCS$ to estimate $P(CS)$ for ordinal optimization.

### 12.4.2 The OCBA Technique

While *ordinal optimization* could significantly reduce the computational cost for identifying the promising region, there is potential for further improvement of performance by intelligently determining the number of simulation replications among different solutions. Intuitively, to ensure a high $P(CS)$ or $APCS$, a larger portion of the computing budget should be allocated to those solutions which are potentially good solutions (or critical solutions) in order to reduce estimator variance. On the other hand, limited computational effort should be expanded on non-critical solutions that have little effect on identifying the good solutions even if they have large variances. Ideally, we want to optimally choose the number of simulation replications for all solutions to maximize simulation efficiency with a given computing budget. In this section, we present a technique called *optimal computing budget allocation* (OCBA) which makes use of this idea. In fact, OCBA is complementary with the NP method. In the NP method, we concentrate sampling effort in the promising regions so that a best solution can be sampled with a higher probability. On the other hand, OCBA balances the simulation accuracy for all sample solutions selected from different regions, so that the overall simulation efficiency can be dramatically improved. We elaborate the OCBA technique in the following.

Let $t_x$ be the number of simulation replications for solution $x$, $H$ be the set of sample solutions to be evaluated and $H = \{x^1, x^2, \cdots, x^h\}$. If simulation is performed on a sequential computer and the difference of computation costs of simulating different solutions is negligible, the total computation cost can be approximated by $\sum_{i=1}^{h} t_{x^i}$. The goal is to choose $t_{x^i}$ for all $x^i$ such that the total computation cost is minimized, subject to the restriction that the ordinal comparison confidence level defined by $APCS$ is greater than some satisfactory level.

$$\min \sum_{i=1}^{h} t_{x^i}$$

s.t. $APCS \geq P^*$.

where $P^*$ is a user-defined confidence level requirement, which corresponds to the stopping criterion in each iteration of the NP method.

Chen et al. (2000b) offer an asymptotic solution, which is summarized in the following theorem.

**Theorem 12.4.** *Assume the simulation is performed on a sequential computer and the difference of computation costs of simulating different solutions is negligible. Given total number of simulation budget $T$ to be allocated to a finite number of competing solutions, as $T \to \infty$, the $APCS$ can be asymptotically maximized when*

$$(a) \frac{t_a}{t_b} \to \frac{s_a}{s_b} \left[ \sum_{x \neq a, x \in H} \left( \frac{\delta_{a,b}^2}{\delta_{a,x}^2} \right) \right]^{1/2}$$

$$(b) \frac{t_x}{t_b} \to \left( \frac{s_x/\delta_{a,x}}{s_b/\delta_{a,b}} \right)$$

*where $a$ is the solution having the largest sample mean, $b$ is the solution having the second largest sample mean, $\delta_{i,j} = \frac{1}{t_i} \sum_{u=1}^{t_i} L(x^i, \xi_u) - \frac{1}{t_j} \sum_{u=1} t_j L(x^j, \xi_u)$, and $s_x$ is the sample standard deviation of solution $x$.*

Using the concepts of ordinal optimization and optimal computing budget allocation, we develop an iterative experimentation procedure for simulation evaluation of the set solution, $H$. The procedure is summarized as follows: We assume total absence of knowledge about any solution considered and any other basis for allocating computing budget at the beginning of the experiment. We first simulate all solutions in $H$ with $t_0$ replications. The sample means and variances can be calculated and then Theorem 12.3 is applied to determine further simulation allocation. More simulation replications are performed based on the allocation. Since the available information about the sample means and variances are approximations, we sequentially update the information and limit the additional computing budget for each iteration to a pre-specified $\Delta$. We repeat the above procedure until $APCS$ is sufficiently high. The algorithm is summarized as follows:

**Algorithm** *Sequential Optimal Computing Budget Allocation (OCBA)*

0. Perform $t_0$ simulation replications for all solutions; $l \leftarrow 0$; $t_{x^1}^l = t_{x^2}^l = \ldots = t_{x^h}^l = t_0$.

1. If $APCS \geq P^*$, stop.
2. Increase the computing budget (i.e., number of additional simulations) by $\Delta$ and compute the new budget allocation, $t_{x^1}^{l+1}$, $t_{x^2}^{l+1}$, ..., $t_{x^h}^{l+1}$, using Theorem 3.
3. Perform additional $\max(0, t_{x^i}^{l+1} - t_{x^i}^{l})$ simulations for solution $x^i$, for each $i$. $l \leftarrow l + 1$. Go to Step 1.

In the above algorithm, $l$ is the iteration number, $APCS$ is used to estimate the ordinal comparison confidence level $P(CS)$, and $P^*$ is a user-defined confidence level requirement. In addition, we need to select the initial number of simulations, $t_0$, and the one-time increment, $\Delta$. Chen et al. (2000a) offers detailed discussions on the selection. It is well understood that $t_0$ cannot be too small as the estimates of the mean and the variance may be very poor, resulting in premature termination of the comparison. A suitable choice for $t_0$ is between 5 and 20 (Law and Kelton 1991, Bechhofer et al. 1995). Also, a large $\Delta$ can result in waste of computation time to obtain an unnecessarily high confidence level. On the other hand, if $\Delta$ is small, we need to compute $APCS$ (in step 1) many times. A suggested choice for $\Delta$ is a number bigger than 5 but smaller than 10% of the simulated solutions. In particular, we set $t_0 = 5$ and $\Delta = 10$ for our numerical experiments reported below.

### 12.4.3 The NP Hybrid Algorithm

We will now describe the hybrid algorithm in detail. In the $k$-th iteration we assume that there is a subregion $\sigma(k) \subseteq X$ of the feasible region that may be considered most promising. Initially we assume no knowledge about the most promising region and let it be $\sigma(0) = X$, that is, we do not use warm-start. The first step is to partition the most promising region into $M_\sigma$ disjoint subregions and aggregate the surrounding region (if any) into one. For example, in Figure 12.2, we first partition the solution space into $M_{\sigma(0)} = 3$ subregions without any surrounding region. Next, if $\sigma(1) = \{2, x_2, x_3\}$ becomes the most promising region, we then partition $\sigma(1)$ into $M_{\sigma(1)} = 2$ subregions and the surrounding region is $X \setminus \sigma(1)$. The second step is to use a random sampling method to generate a set of feasible solutions from each region. As outlined in Chapter 2, this should be done in such a way that each point has a positive probability of being selected. However, as we know, many heuristics may also be incorporated into the sampling step through a weighted sampling scheme, and this will be further illustrate for the resource allocation problem through an example below. The third step is to apply the *ordinal optimization* and *OCBA* technique to rank and select the best from a set of sample solutions, a union of all sample solutions from each disjoint region. The final step is to determine the most promising region for the next iteration. The subregion estimated to have the best promising index becomes the most promising region in the next iteration. The new most promising region is thus nested within the last. By extension, if the surrounding region is found to have the best promising index, the algorithm backtracks to a larger region that contains the best solution. The partitioning continues until singleton regions are obtained and no further partitioning is possible. Thus, each step is the same as in the general NP framework, except in the third step, where both ordinal optimization and the OCBA techniques are incorporated to improve the selection of a sample solution. This can now be summarized in the following algorithm.

**Algorithm** *Hybrid NP/OCBA*

0. **Initialization.** Let $\sigma(0) = X$ and $k = 0$.

1. **Partitioning.** If $|\sigma(k)| > 1$, partition $\sigma(k)$ into $M_{\sigma(k)}$ subregions,

$$\sigma_1(k), \sigma_2(k), \ldots, \sigma_{M_{\sigma(k)}}(k)$$

If $\sigma(k) \neq X$ aggregate the surrounding region into one region,

$$\sigma_{M_{\sigma(k)}+1}(k) = X \setminus \sigma(k)$$

It should be noted that $\sigma_{M_{\sigma(k)}}(k)$ only depends on $\sigma$ but not $k$.

2. **Sampling.** Use a random sampling procedure to select $h_j$ sample points from each subregion $\sigma_j(k)$, $j = 1, 2, \ldots, M_{\sigma(k)} + 1$. Let $D_{\sigma_j(k)}$ denote the set containing the sample solutions in $\sigma_j(k)$, i.e.,

$$D_{\sigma_j(k)} = \{x^{j1}, x^{j2}, .., x^{jh_j}\}. \tag{12.9}$$

3. **Ranking and selection of the best solution via Ordinal Optimization & OCBA.** Use the promising index set performance function $I : \Sigma \to R$ given in (12.5) and its estimation given in (12.6). i.e,

$$I(\sigma) = \max_{x \in \sigma} J(x), \tag{12.10}$$

and

$$\hat{I}(\sigma) = \max_{x \in D_\sigma} \hat{J}(x). \tag{12.11}$$

Then find the index of the most promising region.

$$\hat{j}_k = \arg \max_{j=1,\ldots,M_{\sigma(k)}+1} \hat{I}(\sigma_j(k)). \tag{12.12}$$

This step involves two levels of optimization and is the most time consuming one in our hybrid algorithm. Ordinal optimization and OCBA are applied to efficiently solve the optimization problem in (12.11) and (12.12), and to identify the most promising region. Details are given below.

4. **Backtracking**. If more than one region is equally promising, the tie can be broken arbitrarily (e.g. with equal probability to be selected). If this index corresponds to a region that is a subregion of $\sigma(k)$, then let this be the most promising region in the next iteration. That is let $\sigma(k+1) = \sigma_{\hat{j}_k}(k)$. If $\hat{j}_k$ correspond to the surrounding region, backtrack to a larger region containing $\hat{j}_k$. Let $k = k + 1$. Go back to step 1.

The hybrid algorithm provides an optimization framework for the resource allocation problem. Step 3 is most time consuming. By integrating ordinal optimization and OCBA into Step 3, overall efficiency can be significantly improved. In addition, there is great flexibility to incorporate heuristic methods into Step 2, the random sampling, and Step 3, the estimation of the promising index. In the following section, we consider the buffer allocation problem (Example 1) and present a more detailed, step-by-step description of the algorithm, while incorporating some heuristics in the algorithm.

### 12.4.4 Implementation

We will illustrate the hybrid NP/OO/OCBA by applying it to the buffer resource application problem discussed earlier in this chapter.

### Partitioning

To apply the hybrid NP algorithm to the resource allocation problem, still we need to consider how to partition the solution space. Although the NP method does not limit the way in which we partition, specific strategies employed have impact on the efficiency of the algorithm (see Section 2.2). Through partitioning, if good solutions are clustered together, the NP algorithm will then quickly identify a set of near optimal solutions. For the buffer allocation problem, we can first divide this solution space into $M = 13$ subregions by fixing the first buffer $x_0$ to be $0, 1, \ldots, 12$. We can further partition each such subregion by fixing the second buffer to be any of the remaining numbers. This procedure can be repeated until the singleton region is reached, when all the resources are allocated to the buffers. It can be seen from Figure 12.3 that for the buffer allocation problem, at the same level, each subregion contains a different number of feasible points. This partitioning approach is implemented in the numerical experiments reported below.

### Generating Feasible Sample Solutions

The method used to obtain random samples from each region in each iteration is flexible for the NP algorithm. The only requirement is that each point in a sampling region should have a positive probability of being selected. While uniform sampling scheme works well in most cases, from our experiences, incorporation of a simple heuristic into the sampling scheme can drastically

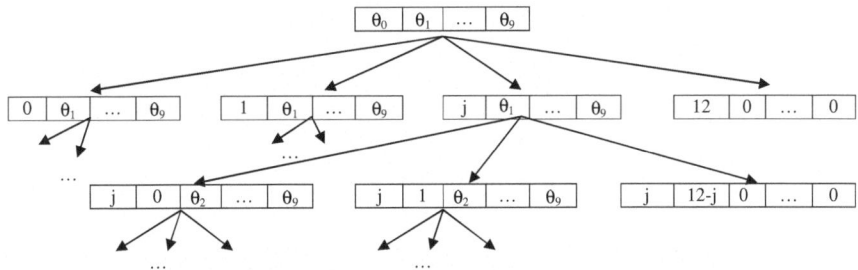

**Fig. 12.3.** Partitioning of the buffer allocation problem.

improve the sampling quality. The sampling scheme used in our numerical testing below is involves a relatively minor change from a uniform distribution. The detailed procedure is given as follows:

Suppose the current sampling region has the form of $\sigma(k) = \{(x_0, \ldots, x_9) | x_0 = \hat{x}_0, \ldots, x_k = \hat{x}_k\}$. This means that the number of buffer units at the first $k$ nodes $(\hat{x}_0, \ldots, \hat{x}_k)$ is fixed. $x_{k+1}, \ldots, x_9$ are to be determined. Generating a random sampling point in $\sigma(k)$ means to randomly select $x_{k+1}, \ldots, x_9$ such that constraint (12.1) $(x_0 + x_1 + x_2 + \ldots + x_9 = 12)$ is satisfied. One way is to sequentially generate a random number for each of $x_i, i = k+1, \ldots, 9$. Let

$$R_i = 12 - \sum_{j=0}^{i-1} x_j.$$

Thus $R_i$ is the number of buffer units that we can allocate to nodes $i$ to 9. It is also an upper bound of $x_i$. To generate a random sample of $x_i$, instead of generating a random number uniformly distributed between 0 and $R_i$, we adopt a very simple heuristic for our sampling scheme. Since we intend to allocate 12 buffer units to 10 nodes, on average, each node can be allocated only 1.2 units. Intuitively, it is very unlikely for a large amount of units to be allocated to a particular node. Conceptually, it is also undesirable for such an allocation to take place, because the system is symmetric. Thus, we favor the allocation with no more than 2 units in each node. For example, we can weigh the probabilities for the numbers between 0 and 2 three times higher than the numbers bigger than 2. This sampling scheme is illustrated in Figure 12.4 with $R_i$ greater than 2.

In cases when $R_i$ is less than or equal to 2, we can just use a uniform distribution for generating a sample for $x_i$. The above procedure is repeated until the last node is allocated ($i = 9$) or all buffer units are used ($R_i = 0$).

## Promising Index using Ordinal Optimization & OCBA

The most promising region in each iteration of the hybrid algorithm is determined by solving two optimization problems defined in (12.11) and (12.12).

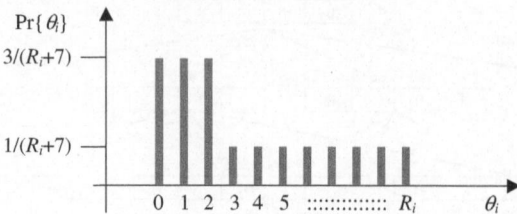

**Fig. 12.4.** Weighted sampling scheme.

(12.11) is used to estimate the promising index (or the best sample solution) in each disjoint region and (12.12) identifies the most promising region. Since we need to only pinpoint the most promising region, we do not have to estimate all the promising indices for all the subregions. Instead, to utilize the OCBA technique, we combine two optimization problem (12.11) and (12.12) into the following optimization problem

$$x^* = \arg\max_{x \in H(k)} \hat{J}(x) \tag{12.13}$$

where $H(k) = \overset{M_{\sigma(k)}+1}{\underset{j=1}{\cup}} D_{\sigma_j(k)}$ is all the sample solutions selected in the $k$-th iteration.

Thus, the most promising region refers to the region to which $x^*$ belongs. The total number of solutions that must be evaluated using simulation in each iteration is $\sum_{j=1}^{M_{\sigma(k)}+1} h_j$ . While $\sum_{j=1}^{M_{\sigma(k)}+1} h_j$ is much smaller than the total number of solutions in the whole solution space, the total simulation cost for all $\sum_{j=1}^{M_{\sigma(k)}+1} h_j$ solutions is still very high. Therefore improving computation efficiency at the most time consuming Step 3 is crucial to the overall efficiency of the hybrid algorithm.

As we mentioned before, the NP method is based on order comparison to determine a promising region. An accurate estimation of the promising index for each region is not necessary. That is, the relative order of promising indices, or the relative order of the considered solutions, is more essential than the value of the promising index itself. This is an ideal situation for the use of ordinal optimization. As shown in Theorem 12.1, if we concentrate on the order comparison, the probability of correctly selecting the solution $x^*$ converges to one exponentially. This means that the probability of correctly identifying the most promising region using ordinal optimization converges to 1 exponentially.

Furthermore, instead of simulating all solutions equally, we apply OCBA to intelligently determine the optimal simulation budget for each solution so that the total simulation cost can be significantly reduced. The algorithm presented in Section 3.3 is applied to the problem in (12.13). As will show in

Section 4, OCBA can achieve a speedup factor of 16 on the top of the use of ordinal optimization.

Note that it is more advantageous if OCBA is applied to the combined problem (12.13) directly than if it is applied to the two-level optimization problems in (12.11) and (12.12) separately. OCBA has the capability to optimally allocate simulation budget by considering the means and variances of all sample solutions from different regions. This way, the information of both solutions and regions can be utilized when allocating simulation budget to achieve a better performance.

**Backtracking**

There are many backtracking rules for the hybrid algorithm. In this paper, we consider the following options for backtracking since they are easy to implement.

*Backtracking Rule I*: Backtracking to the superregion of the current most promising region. That is let:

$$\sigma(k+1) = \begin{cases} \sigma_{\hat{j}_k}(k) \text{if} \hat{j}_k < M_{\sigma(k)} + 1 \\ s(\sigma(k)), \text{otherwise} \end{cases}.$$

We refer to the hybrid algorithm using this backtracking rule as the **Hybrid Algorithm I**. It should be noted that in order to backtrack to the superregion of the current most promising region, we need to keep track of only those regions that lead from the current most promising region back to the entire feasible region. This imposes minimal memory overhead.

*Backtracking Rule II*: Backing tracking to the entire feasible region. That is let:

$$\sigma(k+1) = \begin{cases} \sigma_{\hat{j}_k}(k) \text{if} \hat{j}_k < M_{\sigma(k)} + 1 \\ X, \text{otherwise}. \end{cases}$$

We refer to the hybrid algorithm using this backtracking rule as the **Hybrid Algorithm II**. The difference between these two algorithms can be thought of in terms of long-term memory. If the algorithm II is used, then we can move immediately out of that region in a single transition. For the algorithm I, on the other hand, completely moving out of regions of more depth than one requires more than one transition. Therefore, Algorithm I has long-term memory, but the algorithm II does not. Alternatively, since most of the sampling effort is concentrated in the promising region, we can think of the backtracking feature as a method to change the sampling distribution. Using this interpretation, the Algorithm I only allows for a slight change in the sampling distribution at each iteration, but the algorithm II permits us to make a drastic change.

Other backtracking methods exist that can also change the sampling distribution drastically. Note for example, we can backtrack to any superregion of a singleton region that contains the best solution found in the current iteration. Such a backtracking rule would severely bias the sampling distribution towards the best solution found in each iteration, regardless of it being in the current most promising region or the surrounding region. For example, for the buffer allocation problem, we can use the following backtrack rule that uses the information obtained in the previous iteration.

*Backtracking Rule III*: Assume that

$x^* \equiv (x_0^*, x_1^*, \ldots, x_9^*)$ is the best solution for (12.13) found in the current iteration,

$x^- \equiv (x_0^-, x_1^-, \ldots, x_9^-)$ is the best solution for (12.13) found in the previous iteration.

backtrack to the level that $x^*$ and $x^-$ have the same component at that level and above. That is, backtrack to the partition whose elements are in the form of $(x_0^*, \ldots, x_k^*, x_{k+1}, \ldots, x_9)$ where $k = \max\{i : x_j^* = x_j^-, \text{ for all } j \leq i\}$.

## 12.5 Numerical Results

In this section, we apply the hybrid algorithm to the buffer allocation problem discussed above. Before we report the numerical result of the hybrid algorithm, we first demonstrate how the *ordinal optimization* and OCBA techniques can be applied to a simplified version of the buffer allocation problem. In this simplified version, since the total number of solutions (or solutions) is 210, the ordinal optimization can be directly applied. We show that OCBA can achieve a speedup factor as high as 16 on top of the use of ordinal optimization only. In the next section, we apply the hybrid algorithm to deal with the original buffer allocation problem that has a much larger solution space. We show that a better solution (comparing with the solution obtained from the simplified problem) can be obtained with a reasonable simulation cost. Finally, we apply the hybrid NP algorithm to a more complex and less structured problem with a much bigger solution space. Again, our algorithm converges to the optimal solution with a reasonable simulation cost.

### 12.5.1 A Reduced Problem

Consider the 10-node network presented in Section 12.2 in which the objective is to select a solution with minimum expected time to process the first 100 customers from a same initial state when the system is empty. This is a transient simulation, and $L(x, \xi_i)$ is the simulation result of the $i$-th run. Multiple simulation runs are needed to estimate $E[L(x, \xi)]$ for each $x$. As discussed in Section 2, even for an allocation of 12 buffer units to 10 nodes, there are 293,930 different combinations. While the simulation time for each combination is not very long, the total simulation time for 293,930 solutions are not affordable. To reduce the number of solutions for consideration to a manageable size, observing that the network is symmetric, we then set the

**Table 12.1.** Average total simulation runs for different confidence level of $APCS$ over 10,000 independent simulation experiments. The second column is the needed simulation runs using ordinal optimization only. The third column includes the results for OCBA.

| APCS $\geq P^*$ | Simul. Runs using OO only | Simul. Runs using OO + OCBA | Speedup using OCBA |
|---|---|---|---|
| 60% | 20622 | 2206 | 9.34 |
| 70% | 29211 | 2631 | 11.1 |
| 80% | 48699 | 3522 | 13.8 |
| 90% | 91665 | 5716 | 16.0 |

following three constraints:

$$x_0 = x_1 = x_2 = x_3 \tag{12.14}$$

$$x_4 = x_6 \tag{12.15}$$

$$x_5 = x_7 \tag{12.16}$$

With the above three constraints, the number of solutions considered here is reduced to 210. Since the network is symmetric, we originally anticipated that the optimal solution should satisfy all the above three constraints. This turned out not to be the case after we apply the hybrid algorithm, as we will show later in next subsection.

We first focus on the reduced 210 solutions and apply ordinal optimization and OCBA to this simplified problem. We stop simulation at different confidence levels of $APCS$. Since the required simulation run for a desired confidence level varies from time to time, 10,000 independent experiments are performed and the average simulation runs are compared. Table 1 compares the averages of the required simulation runs for different confidence levels for ordinal optimization and the integration of ordinal optimization and OCBA.

From the Table 12.1, we observe that a higher computing budget can lend readily to a higher $APCS$. Using the OCBA scheme, however, significantly reduces the computation cost for a desired level of $APCS$. Furthermore, the speedup of using OCBA increases, as the desired confidence level becomes higher. The speedup factor is as high as 16 as $P^* = 90\%$. We anticipate that the observed savings will become even more significant when $P^*$ is higher, since we will have more flexibility to manipulate the budget allocation.

In order to have a better idea about the optimal solution, we conduct a simulation experiment for all 210 solutions with $P^* = 99.999\%$. The best solution we obtained is $[x_0, x_1, x_2, \ldots, x_9] = [1, 1, 1, 1, 2, 1, 2, 1, 1, 1]$. We will show that the proposed hybrid algorithm can obtain a better solution with a reasonable simulation cost in the next subsection.

**Table 12.2.** The promising region at each iteration.

| Iteration | Action | Promising Region | | | | | | | | | |
|---|---|---|---|---|---|---|---|---|---|---|---|
| | | $x_0$ | $x_1$ | $x_2$ | $x_3$ | $x_4$ | $x_5$ | $x_6$ | $x_7$ | $x_8$ | $x_9$ |
| 0 | Sampling | | | | | | | | | | |
| 1 | Partition | 1 | | | | | | | | | |
| 2 | Partition | 1 | 1 | | | | | | | | |
| 3 | Backtracking | 2 | | | | | | | | | |
| 4 | Partition | 2 | 1 | | | | | | | | |
| 5 | Partition | 2 | 1 | | | | | | | | |
| 6 | Partition | 2 | 1 | 1 | 1 | | | | | | |
| 7 | Partition | 2 | 1 | 1 | 1 | | | | | | |
| 8 | Partition | 2 | 1 | 1 | 1 | 0 | | | | | |
| 9 | Partition | 2 | 1 | 1 | 1 | 0 | | 2 | | | |
| 10 | Partition | 2 | 1 | 1 | 1 | 0 | 2 | 1 | | | |
| 11 | Backtracking | 2 | 1 | 1 | 1 | 2 | | | | | |
| 12 | Partition | 2 | 1 | 1 | 1 | 2 | 1 | | | | |
| 13 | Partition | 2 | 1 | 1 | 1 | 2 | 1 | 2 | | | |
| 14 | Partition | 2 | 1 | 1 | 1 | 2 | 1 | 2 | 1 | | |
| 15 | Stop | 2 | 1 | 1 | 1 | 2 | 1 | 2 | 1 | 0 | |

### 12.5.2 The Original Resource Allocation Problem

In this subsection we apply the hybrid algorithm to the original 10-node network. Note that the problem considered here has 293,930 different solutions that are dramatically bigger than the 210 considered in the reduced problem.

In each iteration, we follow the partition rules described in above and illustrated in Figure 12.3. Namely, we first divide this solution space into $M = 13$ subregions by fixing the first buffer $x_0$ to be $0, 1, \ldots, 12$. We then further partition each such subregion by fixing the second buffer to be any of the remaining numbers. We randomly sample 45 solutions from the promising region, and 105 solutions from the surrounding region, creating a total of 150 solutions for consideration in an iteration (i.e., $|H(k)| = 150$). The stopping criterion is set at the point where the confidence level of identifying the best in the 150 solutions is no less than 90%, i.e., $APCS > 90\%$. In the OCBA algorithm, we set $t_0 = 5$ and $\Delta = 10$. In order to improve the quality of our sampling solutions, we adopt the simple heuristic illustrated in Figure 12.4 for our sampling scheme. Table 12.2 shows how our algorithm evolves by giving the promising region at each iteration. Note that $x_9$ is determined as soon as $x_0, x_1, x_2, \ldots$, and $x_8$ are fixed at iteration 15, because the total number of buffer units is given.

From Table 12.2, we can see that after 15 iterations, our algorithm converges to a solution $[x_0, x_1, x_2, \ldots, x_9] = [2, 1, 1, 1, 2, 1, 2, 1, 0, 1]$. It turns out that this solution is better than the solution we found in the previous section. Obviously, this solution does not satisfy the symmetric constraints in

**Table 12.3.** Different queue disciplines considered at the four arriving nodes.

| Queue discipline | | Descriptions |
|---|---|---|
| A | FIFO | First In First Out |
| B | LIFO | Last In First Out |
| C | C1F | Class 1 customers have higher priority than Class 2 customers |
| D | C2F | Class 2 customers have higher priority than Class 1 customers |

(12.14,12.15,12.16). If we had relied simply on the reduced problem, we would never have uncovered this solution. The total number of simulation runs to converge to this solution is 17,379, which is about only 3 times bigger than the cost needed for the reduced problem discussed above. Given that the solution space is much bigger ($293, 930/210 \approx 1400$ bigger), the time saving is highly significant.

It is interesting to see that the optimal solution is not symmetric, although the structure of the network is. We guess this is due to the integer constraint of the buffer size. If the integer constraint is relaxed (for example, 0.8 buffer unit is allowed), then the optimal solution should be symmetric.

In our simulation, one arrival event and three departure events must be generated in order to simulate one customer. 100 customers are simulated in one run based on the performance measure requirement. Thus, the total number of events generated for converging to the optimal solution is about $4 \times 100 \times 17, 379 = 6, 951, 600$. While the number of needed evens is problem-specific, the CPU time on a SUN Sparc 20 workstation is less than one minute. Please note that our algorithm does not utilize the property of the problem structure. Some other algorithms, which utilize the structure information to direct the search direction, such as Cassandras et al. (1998) may find an optimal solution with an even lower computation cost. However, our algorithm can be effectively applied to other general problems without knowing much about the problem structure. We present one example in the next subsection.

### 12.5.3 A More Complex and Less Structured Problem

In order to show that our algorithm is equally applicable to less structured problems, we consider a more complex problem. In each of the arriving nodes (nodes 0-3) of the 10-node network, class 1 and class 2 customers wait in the same queue when the server is busy. Whenever the server is available for serving customers in the queue, the two types of customers have to compete for a same server. In addition to the FIFO considered in earlier examples, three more queue disciplines are included herein for each of these four arriving nodes, which are depicted in Table 3.

With the inclusion of different queue disciplines at each of the four arriving nodes, the total number of solutions is increased to $293, 930 \times 44 = 75, 246, 080$. Furthermore, there is no obvious relationship between any two disciplines in terms of their impact on the performance. Traditional gradient-search methods therefore can not be directly applied to such a problem.

Since it is impossible to exhaustively search the whole solution space, we carefully select a performance measure so that we know what the optimal solution is and then we can check how well our algorithm works. In this problem, the objective is to select a solution with minimum expected time to process the first 100 arriving customers from a same initial state. It is clear that the best queue discipline is FIFO, because other disciplines may change the order of customer service and so delay the completion of the targeted 100 customers.

All the settings in our algorithm are the same as those in the previous section. Our algorithm converges to a solution $[x_0, x_1, x_2, \ldots, x_9] = [2, 1, 1, 1, 2, 1, 2, 1, 0, 1]$, which should be the optimal solution based on our observation. The total number of simulation runs to converge to this solution is 19,281, which is not much higher than the cost needed for the original problem. Given that the solution space is much bigger (256 times bigger), our algorithm is efficient when handling big and complex problems.

## 12.6 Conclusions

In this chapter we consider a hybrid NP algorithm to solve a stochastic discrete resource allocation optimization. The problem considered here is very difficult to solve since it is typically susceptible to noisy estimation of the performance function and exponentially increased solution space. As described in Chapter 3, the NP method is very applicable to such problems, and here we incorporated the paradigm of *ordinal optimization* and an efficient ranking and selection technique called *optimal computing budget allocation* (OCBA) into the NP framework. Specifically, the NP method provides the overall framework, structuring of the search space through partitioning, and uses random sampling to generate feasible solution. Ordinal optimization and the OCBA technique are then incorporated to improve the comparison of sample solutions that are generated. This hybrid NP algorithm can be applied to a wide variety of resource allocation problems, and its efficiency was been demonstrated through a numerical example.

# References

1. Aardal, K., 1998, "Capacitated Facility Location: Separation Algorithms and Computational Experience," Mathematical Programming **81**, 149–175, (1998).
2. Adiri, I. and Yehudai, Z. (1987) Scheduling on Machines with Variable Service Rates. *Computers and Operations Research*, **14**, 289–97.
3. Aha, D.W., R.L. Bankert. 1996. A comparative evaluation of sequential feature selection algorithms. D. Fisher and J.-H. Lenz, eds. *Artificial Intelligence and Statistics V*. Springer-Verlag, New York.
4. T. Altiok and S. Stidam (1983), The allocation of Inter-stage Buffer Capacities in Production Lines, IIE Transactions, Vol. 15, No. 4, 292–299.
5. A. Andijani and M. Anwarul (1997), Manufacturing Blocking Discipline: A Multi-Criteria Approach for Buffer Allocation, International Journal of Production Economics, Vol. 51, 155–163
6. Andradóttir, S. 1995. "A Method for Discrete Stochastic Optimization," *Management Science*, 41, 1946–1961.
7. Andradóttir, S. 1996. "A Global Search Method for Discrete Stochastic Optimization," *SIAM Journal on Optimization* **6**, 513–530.
8. Balakrishnan, P.V. and V.S. Jacob, "Genetic Algorithms for Product Design," *Management Science* **42**, 1105–1117 (1996).
9. —— and ——, "Triangulation in Decision Support Systems: Algorithms for Product Design," *Decision Support Systems* **14**, 313–327 (1995).
10. Basu, A. 1998. Perspectives on operations research in data and knowledge management. *European Journal of Operational Research* **111** 1–14.
11. Beasley, J., 1993, Lagrangian Heuristics for Location Problems, European Journal of Operational Research 65, pp 383–399.
12. Blake, C.L., C.J. Merz. 1998. *UCI repository of machine learning databases.* http://www.ics .uci.edu/~mlearn/MLRepository.html. Department of Information and Computer Science, University of California, Irvine, CA.
13. Boxma, O.J., Rinnooy Kan, A.H.G. and Van Vliet, M. (1990) Machine Allocation Problems in Manufacturing Networks. *European Journal of Operational Research*, **45**, 47–54.
14. Bradley, P.S., U.M. Fayyad, O.L. Mangasarian. 1999. Mathematical programming for data mining: formulations and challenges. *INFORMS Journal on Computing* **11** 217–238.

15. Bradley, P.S., O.L. Mangasarian, W.N. Street. 1998. Feature selection via mathematical programming. *INFORMS Journal on Computing* **10** 209–217.

16. Bramel, J. and Simchi-Levi, D., 1997, The Logic of Logistics: theory, algorithms, and applications for logistics management, Springer Series in Operations Research.

17. J. A. Buzzocott (1967), Automatic Transfer Lines with Buffer Stocks, International Journal of Production Research, Vol. 6, 183–200

18. Caruana, R., D. Freitag. 1994. Greedy attribute selection. *Proceedings of the Eleventh International Conference on Machine Learning.* Morgan Kaufmann. New Brunswick, NJ. 28–36.

19. Cassandras, C.G., L. Dai, and C.G. Panayiotou. 1998. "Ordinal Optimization for a Class of Deterministic and Stochastic Discrete Resource Allocation Problems," *IEEE Transactions on Automatic Control,* 43, 881–900.

20. Cheng, T.C.E. and Sin, C.C.S. (1990) A State-of-the-Art Review of Parallel-Machine Scheduling Research. *European Journal of Operational Research,* **47**, 271–92.

21. R. W. Conway et al (1988), The Role of Work-In-Process Inventory in Serial Production Lines, Operations Research, Vol. 36, 229–241

22. Crainic, T., Toulouse, M., and Gendreau, M., 1996, Parallel asynchronous tabu search for multicommodity location-allocation with balancing requirements, Annals of Operations Research 63, pp277–299

23. Dai, L. 1996. "Convergence Properties of Ordinal Comparison in the Simulation of Discrete Event Dynamic Systems," *Journal of Optimization Theory and Applications,* 91, 363–388

24. Dai, L. and C.H. Chen. 1997. "Rates of Convergence of Ordinal Comparison for Dependent Discrete Event Dynamic Systems," *Journal of Optimization Theory and Applications,* 94, 29–54.

25. Delaney, R., 2003, 14th Annual State Of Logistics Report, June 2, National Press Club Washington, D. C.

26. Y.Dallery, R. David and X.L.Xie (1988), An efficient Decomposition method for the approximate evaluation of tandem queues with finite storage space and blocking, IIE Transactions, Vol.20, 280–283

27. Y.Dallery, R. David and X.L.Xie (1989), Approximate Analysis of Transfer Lines with Unreliable Machines and Finite Buffers, IEEE Transactions on Automatic Control, Vol. 34, No.9, 943–953

28. Daniels, R.L. and Mazzola, J.B. (1994) Flow Shop Scheduling with Resource Flexibility. *Operations Research,* **42**, 504–22.

29. Daniels, R.L., Hoopes, B.J. and Mazzola, J.B. (1996) Scheduling Parallel Manufacturing Cells with Resource Flexibility. *Management Science,* **42**, 1260–76.

30. Daniels, R.L. and Sarin, R.K. (1989) Single Machine Scheduling with Controllable Processing Times and Number of Jobs Tardy. *Operations Research,* **37**, 981–4.

31. M. B. M. DE Koster (1987), Estimation of Line Efficiency by Aggregation, International Journal of Production Research, Vol. 25, 615–626.

32. Delmaire, H., Daz, J., and Fernndez, E., 1999, Reactive GRASP and tabu search based heuristics for the single source capacitated plant location problem, INFOR, Canadian Journal of Operational Research and Information Processing 37(3), pp194–225

33. Dobson, G. and S. Kalish, "Heuristics for Pricing and Positioning a Product Line Using Conjoint and Cost Data," *Management Science* **39**, 160–175 (1993).

34. D'Souza, W., Meyer, R., Naqvi, S., and Shi, L., 2003, Beam orientation optimization in imrt using single beam characteristics and mixed-integer formulations, AAPM Annual Meeting, San Diego

35. Fayyad, U.M. K.B. Irani. 1993. Multi-interval discretisation of continuous-valued attributes. *Proceedings of the Thirteenth International Joint Conference on Artificial Intelligence.* Chambery, France. 1022–1027.

36. Fisher, M., 1981, The Lagrangian Relaxation Method for Solving Integer Programming Problems, Management Sciences 27, pp1–18

37. Frenk, H., Labbé, M., Van Vliet, M. and Zhang, S. (1994) Improved Algorithms for Machine Allocation in Manufacturing Systems. *Operations Research*, **42**, 523–30.

38. C.D.Geiger, K.G.Kempt and U.Uzsoy (1997), A Tabu Search Approach to Scheduling an Automated Wet Etch Station, Journal of Manufacturing Systems, Vol. 16, No.2, 102–116

39. Geofrion, A. and Graves, G., 1974, Multicommodity Distribution System Design by Benders Decomposition, Management Science 20, pp 822–844.

40. S. B. Gershwin (1987), An Efficient Decomposition Method for the Approximate Evaluation of Tandem Queues with Finite Storage Space and Blocking, Operations Research, Vol. 35, 291–305.

41. S. B. Gershwin (1994), Manufacturing Systems Engineering, Prentice Hall, New Jersey

42. Glover (1986), Future Paths for Integer Programming and Links to Artificial Intelligence, Computers and Operations Research, Vol. 13, 533–549

43. F. Glover and M. Laguna(1997), Tabu Search, Kluwer Academic Publishers, Boston

44. Gong, W.-B., Y.-C. Ho, and W. Zhai. 1999. "Stochastic Comparison Algorithm for Discrete Optimization with Estimation," *SIAM Journal on Optimization,* 10, 384–404.

45. Green, P.E., J.D. Carrol, and S.M. Goldberg, "A General Approach to Product Design Optimization via Conjoint Analysis," *Journal of Marketing* **45**, 38–48 (1981).

46. ⸻ and A.M. Krieger, "Models and Heuristics for Product Line Selection," *Marketing Science* **4**, 1–19 (1985).

47. ⸻ and ⸻, "Recent Contributions to Optimal Product Positioning and Buyer Segmentation," *European Journal of Operations Research* **41**, 127–141 (1989).

48. ⸻ and ⸻, "A simple heuristic for selecting 'good'products in conjoint analysis," *J. Advances in Management Science* **5**, R. Schultz (ed.), JAI Press, Greenwich, CT (1987).

49. ⸻ and ⸻, "An application of a product positioning model to pharmaceutical products," *Marketing Science* **11**: 117–132 (1992).

50. ⸻ and V. Srinivasan, "Conjoint Analysis in Consumer Research: New Developments and Directions," *Journal of Marketing* **54**, 3–19 (1990).

51. Hall, M.A. 2000. Correlation-based feature selection for discrete and numeric class machine learning. *Proceedings of the Seventeenth International Conference on Machine Learning.* Stanford University, Stanford, CA. Morgan Kaufmann. 359–366.

52. J. H. Harris and S. G. Powell (1999), An Algorithm for Optimal Buffer Placement in Reliable Serial Lines, Vol. 31, 287–302.

53. F. S. Hillier and R. W. Boling (1966), The Effect of Some Design Factors on the Efficiency of Production Lines with Variable Operation Times, Journal of Industrial Engineering, Vol. 17, 651–658

54. F. S. Hillier and K. C. So and R. W. Boling (1993), Notes: Toward Characterizing the Optimal Allocation of Storage Space in Production Line Systems with Variable Processing Times, Management Sciences, Vol. 39, 126–133

55. Hindi, K. and Basta, T., 1994, Computationally Efficient Solution of a Multiproduct, Two-Stage Distribution-Location Problem, Journal of the Operational Research Society 45, pp 1316–1323.

56. Hindi, K., Basta, T., and Pienkosz, K., 1998, Efficient solution of a multi-commodity, two-stage distribution problem with constraints on assignment of customers to distribution centers, International Transactions in Operations Research 5(6), pp 519–528.

57. Ho, Y.-C. 1997. "On the Numerical Solutions to Stochastic Optimization Problems," *IEEE Transactions on Automatic Control,* 42, 727–729.

58. Y. C. Ho, M.A. Eyleer and T. T. Chien (1979), A Gradient Technique for General Buffer Storage Design in a Production Line, International Journal of Production Research, Vol. 17, 557–580

59. Ho, Y.-C., R.S. Sreenivas, and P. Vakili. 1992. "Ordinal Optimization of DEDS," *Discrete Event Dynamic Systems: Theory and Applications,* 2, 61–88.

60. M. A. Jafari and J. G. Shanthikumar (1987), Exact and Approximate solutions to Two-Stage transfer Lines with general Uptime and Downtime Distribution, IIE Transactions, Vol. 19, 412–420

61. Karabati, S. and Kouvelis, P. (1997) Flow-line Scheduling Problem with Controllable Processing Times. *IIE Transactions,* **29**, 1–15.

62. Karabati, S., Kouvelis, P. and Yu, G. (1995) The Discrete Resource Allocation Problem in Flow Lines. *Management Science,* **41**, 1417–30.

63. Kekre, S. and K. Srinivasan, "Broad Product Line: A Necessity to Achieve Success?" *Management Science* **36**: 1216–1231 (1990).

64. Kim, Y.S., W.N. Street, F. Menczer. 2000. Feature selection in unsupervised learning via evolutionary search. *Proceedings of the 6th ACM SIGKDD International Conference on Knowledge Discovery and Data Mining.* Boston, MA. 365–369.

65. Klincewicz, J. and Luss, H., 1986, A Lagrangian Relaxation Heuristic for Capacitated Facility Location with Single-Source Constraints, Journal of Operational Research Society 37, pp 495–500.

66. Klose, A., 2000, A Lagrangean relax-and-cut approach for the two-stage capacitated facility location problem, European Journal of Operational Research 126, pp 408–421.

67. Klose, A. and Drexl, A., 2003, Facility location models for distribution system design, European Journal of Operational Research, In Press, Corrected Proof, Available online 15 January 2004.

68. Kohli, R. and R. Krishnamurti, "A Heuristic Approach to Product Design," *Management Science* **33**, 1523–1533 (1987).

69. —— and ——, "Optimal Product Design Using Conjoint Analysis: Computational Complexity and Algorithms," *European Journal of Operations Research* **40**, 186–195 (1989).

70. —— and Sukumar, "Heuristics for Product-Line Design using Conjoint Analysis," *Management Science* **35**: 1464–1478 (1990).

71. Lee, C., 1993, A Cross Decomposition Algorithm for A Multiproduct-Multitype Facility Location Problem, Computers and Operations Research 20, pp527–540

72. Lee, L.H., T.W.E. Lau, and Y.-C. Ho. 1999. "Explanation of Goal Softening in Ordinal Optimization," *IEEE Transactions on Automatic Control,* 44, 94–99.

73. M. Litzkow, M., and M. Mutka. 1988. "Condor - A Hunter of Idle Workstations," *Proceedings of the 8th International Conference of Distributed Computing Systems,* pages 104–111.

74. C. M. Liu and C.I. Lin (1994), Performance Evaluation of Unbalanced Serial Production Lines, International Journal of Production Research, Vol. 12, 2897–2914

75. Liu, H., and H. Motoda. 1998. *Feature Extraction, Construction and Selection: A Data Mining Perspective,* Kluwer Academic Publishers. Boston, MA.

76. Lovsz, L., 1996, Randomized Algorithms in Combinatorial Optimization, Combinatorial Optimization, DIMACS Series in Discrete Mathematics and Theoretical Computer Science, American Mathematical Society pp153–179

77. M. Mascolo, R. David and Y. Dallery (1991), Modelling and Analysis of Assembly Systems with Unreliable Machines and Finite Buffers, IIE Transactions, Vol. 23, No. 4, 315–331

78. Mathar, R. and Niessen, T., 2000, Optimum positioning of base stations for cellular radio networks, Wireless Networks 6(6), pp421–428

79. Mazzola, J. and Neebe, A., 1999, Lagrangian-relaxation-based solution procedures for a multiproduct capacitated facility location problem with choice of facility type, European Journal of Operational Research 115, pp285–299

80. Modrzejewski, M. 1993. Feature selection using rough sets theory. P.B. Brazdil, ed., *Proceedings of the European Conference on Machine Learning.* Vienna, Austria. 213–226.

81. Mirchandani, P. and Francis, R., 1990, Discrete Location Theory, John Wiley and Sons, Inc.

82. Mitchell, J., 2000, Branch-and-Cut Algorithms for Combinatorial Optimization Problems, to appear in the Handbook of Applied Optimization, Oxford University Press

83. Nair, S.K., L.S. Thakur, and K. Wen, "Near Optimal Solutions for Product Line Design and Selection: Beam Search Heuristics," *Management Science* **41**, 767–785 (1995).

84. Narandra, P.M., K. Fukunaga. 1977. A branch and bound algorithm for feature subset selection. *IEEE Transactions on Computers* **26** 917–922.

85. Neebe, A. and Rao, M., 1983, An Algorithm for the Fixed-Charge Assigning Users to Sources Problem, Journal of Operational Research Society 34, pp1107–1113

86. Norking, W.I., Y.M. Ermoliev, and A. Ruszczyński. 1998. "On Optimal Allocation of Indivisables Under Uncertainty," *Operations Research,* 46, 381–395.

87. Ólafsson, S. 1999. Iterative ranking and selection for large-scale optimization. *Proceedings of the 1999 Winter Simulation Conference.* Phoenix, AZ. 479–485.

88. Ólafsson, S. and L. Shi. 1999. "Optimization via Adaptive Sampling and Regenerative Simulation," in P.A. Farrington, H.B. Nembhard, D.T. Sturrock, and G.W. Evans (eds.), *Proceedings of the 1999 Winter Simulation Conference,* 666–672.

89. Ólafsson, S. and L. Shi, "A Method for Scheduling in Parallel Manufacturing Systems with Flexible Resources," *IIE Transactions*, **32**, 135–142, (2000).

90. Ólafsson, S., L. Shi. 2002. Ordinal comparison via the nested partitions method. *Journal of Discrete Event Dynamic Systems* **12** 211–239.

91. Pinedo, M. (1995) *Scheduling: Theory, Algorithms, and Systems.* Prentice-Hall, Englewood Cliffs.

92. Pirkul, H. and Jayaraman, V., 1996, Production, Transportation, and Distribution Planning in a Multi-Commodity Tri-Echelon System, Transportation Sciences 30(4), pp291–302.

93. S. G. Powell and D. F. Pyke (1998), Buffering Unbalanced Assembly Systems, IIE Transactions, Vol. 30, 55–65.

94. Quinlan, J.R. 1986. Induction of decision trees. *Machine Learning* **1** 81–106.

95. N. P. Rao (1976), A Generalization of the Bowl Phenomenon in Series Production Systems, International Journal of Production Research, Vol. 14, No. 4, 437–443

96. C.R.Reeves (Editors)(1993), Modern Heuristic Techniques for Combinatorial Problems, Blackwell Scientific publications, Oxford

97. Santos, C., Zhu, X., and Crowder, H., 2002, A Mathematical Optimization Approach for Resource Allocation in Large Scale Data Centers, Technical Report HPL-2002-64(R.1), Intelligent Enterprise Technologies Laboratory HP Laboratories Palo Alto, http://www.hpl.hp.com/techreports/2002/HPL-2002-64R1.pdf

98. Shi, L. and S. Ólafsson. 2000a. "Nested Partitions Method for Global Optimization," *Operations Research*, 48, 390–407.

99. Shi, L. and S. Ólafsson. 2000b. "Nested Partitions Method for Stochastic Optimization," *Methodology and Computing in Applied Probability*, 2, 271–291.

100. ——, S. Ólafsson, and N. Sun, "New Parallel Randomized Algorithms for the Traveling Salesman Problem," *Computers & Operations Research* **26**, 371–394 (1999).

101. L. Shi and S. Olafasson and Q. Qun (1999), A New Hybrid Optimization Algorithm, Computer and Industrial Engineering, Vol. 36, 409–426.

102. Shih, Y.-S. 1999. Families of splitting criteria for classification trees. *Statistics and Computing* **9** 309–315.

103. Simchi-Levi, D., Kaminsky, P., and Simchi-Levi, E., 2000, Designing and Managing The Supply Chain, Irwin McGraw-Hill

104. Skalak, D. 1994. Prototype and feature selection by sampling and random mutation hill climbing algorithms. *Proceedings of the Eleventh International Machine Learning Conference.* Morgan Kaufmann, New Brunswick, NJ. 293–301.

105. V.J.R.Smith et al (1996), Modern Heuristic Search Methods, John Wiley and Sons, Chichester

106. So, K.C. (1990) Some Heuristics for Scheduling Jobs on Parallel Machines with Setups. *Management Science*, **36**, 467–75.

107. K. C. So (1997), Optimal Buffer Allocation Strategy for Minimizing Work-in-Process Inventory in Unpaced Production Lines, IIE Transactions, Vol. 29, 81–88

108. Tang, Z.B. 1994. "Adaptive Partitioned Random Search to Global Optimization," *IEEE Transactions on Automatic Control*, 39, 2235–2244.

109. Trick, M.A. (1994) Scheduling Multiple Variable-Speed Machines. *Operations Research*, **42**, 234–48.

110. Van Wassenhove, L.H. and Baker, K.R. (1982) A Bicriterion Approach to Time Cost Trade-Offs in Sequencing. *European Journal of Operational Research*, **11**, 48–54.
111. Vickson, R.G. (1980) Choosing the Job Sequence and Processing Times to Minimize Processing Plus Flow Cost on a Single Machine. *Operations Research*, **28**, 1115–67.
112. G. A. Vouros and H. T. Papadopoulos (1998), Buffer Allocation in Unreliable Production Lines Using a Knowledge Based System, Computers and Operations Research, Vol. 25, No.12, 1055–1067
113. Witten I.H., E. Frank, L. Trigg, M. Hall, G. Holmes, S.J. Cunningham. 1999. Weka: Practical machine learning tools and techniques with Java implementations. N. Kasabov, K. Ko, eds., *Proceedings of the ICONIP/ANZIIS/ANNES'99 International Workshop: Emerging Knowledge Engineering and Connectionist-Based Information Systems*. Dunedin, New Zealand. 192–196.
114. Wolsey, L., 1998, Integer Programming, John Wiley & Sons, Inc.
115. Xie, X.L. 1997. "Dynamics and Convergence Rate of Ordinal Comparison of Stochastic Discrete Event Systems," *IEEE Transactions on Automatic Control*, 42, 586–590.
116. H. Yamashita and T. Altiok (1998), Buffer Capacity Allocation for a Desired Throughput in Production Lines, IIE Transactions, Vol. 30, 883–891.
117. Yan, D. and H. Mukai. 1992. "Stochastic Discrete Optimization," *SIAM Journal Control and Optimization*, 30, 594–612.
118. Yang, J., V. Honavar. 1998. Feature subset selection using a genetic algorithm. H. Motada, H. Liu, eds, *Feature Selection, Construction, and Subset Selection: A Data Mining Perspective*. Kluwer, New York. 117–136.
119. Zufryden, F., "A Conjoint-Measurement-Based Approach to Optimal New Product Design and Market Segmentation," in *Analytical approaches to product and market planning*, A.D. Shocker (Ed.), Marketing Science Institute, Cambridge, MA (1977).

# Index

*\* A list of the more recent publications in the series is at the front of the book\**